Organized Crime: Culture, Markets and Policies

STUDIES IN ORGANIZED CRIME

Volume 7

Series Editor:
Frank Bovenkerk, *University of Utrecht, Willem Pompe Institute, The Netherlands*

Organized Crime: Culture, Markets and Policies

Editors

Dina Siegel

VU University of Amsterdam

Amsterdam, The Netherlands

and

Hans Nelen

Maastricht University

Maastricht, The Netherlands

Dina Siegel
Department of Criminology
VU University of Amsterdam
Amsterdam, The Netherlands
d.siegel@rechten.vu.nl

Hans Nelen
Department of Criminal Law and Criminology
Maastricht University
Maastricht, The Netherlands
hans.nelen@strafr.unimaas.nl

ISBN 978-0-387-09710-7   ISBN 978-0-387-74733-0 (eBook)

Acknowledgements

In 2001, The Departments of Criminology of The VU University of Amsterdam (VU) and Erasmus Universiteit Rotterdam (EUR) founded the Centre for Information and Research on Organised Crime (CIROC). One of the goals of this network organisation, later joint by the Research and Documentation Centre of the Dutch Ministry of Justice (WODC) and Maastricht University (UM), is to build a bridge between social scientists on the one hand, and policy makers and law enforcement officials on the other hand. In order to achieve this goal, during the last 6 years every 3 months a seminar has been organised in Amsterdam on a specific topic with regard to organised crime and its containment.

In 2003, the most relevant papers that had been presented at CIROC-seminars in 2002 and 2003 were published in the book *Global Organized crime. Trends and developments* (Siegel, van de Bunt, Zaitch, 2003). The 'experiment' turned out to be successful, which encouraged us to continue this initiative. Between 2004 and March 2007, 12 new seminars on different issues related to organized crime activities were organized. This book includes a new selection of scientific papers, based on the presentations during these seminars.

We are grateful to the authors of this book, firstly for their presentations during the CIROC seminars, and secondly for doing their best to finish their articles on time for this publication. We also would like to thank the members of the Scientific Board of CIROC, who reviewed, commented and made the most useful advices to the authors and editors of this book (see appendix for an overview of the members of the scientific board.)

A special mention is due to the following members of CIROC for their valuable contribution to the organization of the seminars: Wasja Rijs, Anna Tulleners, Henk van de Bunt, Edward Kleemans, and Roelof Jan Bokhorst. We also express our gratitude to Krista Huisman and Hendrik Jan Schwencke for their editorial assistance.

Contents

Introduction

Dina Siegel and Hans Nelen

The term 'global organized crime' has been in use in criminology since the mid 1990s. Even more general and abstract than its daughter-terms (transnational or cross-border organized crime), 'global organized crime' seems to embrace the activities of criminal groups and networks all around the planet, leaving no geographical space untouched. The term appears to cover the geographical as well as the historical domain: 'global' has taken on the meaning of 'forever and ever'.

Global organized crime is also associatively linked with 'globalisation'. The social construction of both terms in scientific discourse is in itself an interesting theme. But perhaps even more interesting, especially for academics trying to conduct empirical research in this area, is the analysis of the symbolic and practical meaning of these concepts. How should criminologists study globalisation in general and global organized crime in particular? Which instruments and 'theoretical luggage' do they have in order to conduct this kind of research? The aim of this book is not to formulate simple, straightforward answers to these questions, but rather to give an overview of contemporary criminological research combining international, national and local dimensions of specific organized crime problems. The term global organized crime will hardly be used in this respect. In other social sciences, such as anthropology, there is a tendency to get rid of vague and abstract terms which can only serve to confuse our understanding. In our opinion, criminology should follow this initiative. Criminological studies on organized crime should take into consideration the situational context that is relevant for its origins, as well as its development. The emphasis should be on specific local situations, on people and goods, offenders and victims, their interactions and their symbolic presentation.

The contributions in the first part of this book on *historical transformations* underline the fact that the nature and volume of organized crime – as well as its containment – depend on the social, economic, and cultural context in which organized crime develops and becomes manifest. Anton Blok describes the evolution of the Sicilian mafia throughout the twentieth century, while Letizia Paoli focuses on the containment and prevention of organized crime in the southern Italian context. Paoli argues that mafia groups' current decline is primarily the result of the intensification of law enforcement repression since the early 1990s. Emanuel Marx analyses how the hashish trade has become a major and seemingly permanent feature of

D. Siegel and H. Nelen (eds.), *Organized Crime: Culture, Markets and Policies.*
© Springer 2008

the South Sinai Bedouin economy. In this context, he examines various aspects of daily life in Sinai. The Bedouin not only consider the drugs trade a legitimate economic enterprise, but the requirements for this trade also fit into their social realities and practices.

The central premise of this book that we need more in-depth and empirically founded knowledge on organized crime in various situational contexts, rather than trying to predict how organized crime will develop throughout the world, is also reflected in the second part on *transnational flows*. Sheldon Zhang and Samuel Pineda argue that official corruption may be just as important a contributor in enabling human trafficking activities as any other socio-cultural factor. The contribution of Stefano Becucci also deals with human trafficking, but this author puts emphasis on recent developments within the sex exploitation market in Italy. Becucci's main argument is that foreign criminal associations tend to adopt different organizational models in recruiting 'sex slaves' in comparison with their Italian counterparts. Tihomir Bezlov and Philip Gounev examine the vehicle theft market in Bulgaria. Their contribution not only clarifies the nature and extent of this market, but also relates changes in society (and institutions) with the dynamics of criminal markets.

Many scholars have pointed out that there is a strong symbiotic relationship between organized crime and the legitimate environment in which it flourishes. The metaphor of a clean, innocent society that is being threatened by the evil of organized crime is definitely out of date. The third part of this book focuses on the *intertwinement of illegitimate and legitimate activities*. Dina Siegel reveals the latest developments in the diamond sector in Antwerp and shows why specific features of this sector are of interest to organized crime groups in the context of smuggling, money laundering, organized robberies and fraud. Tim Boekhout van Solinge also moves away from classical organized crime research topics and focuses on the trade in tropical timber. His description and analysis of the timber industry is a good example of how legitimate and illegitimate economic activities are interwoven.

Criminological research indicates that lawyers, notaries, real estate agents, tax consultants, accountants, bankers and other professionals sometimes play a delicate role in shielding crimes and criminal proceeds from the authorities. Henk van de Bunt focuses on the role of Hawala bankers in the transfer of proceeds from organized crime. His main argument is that the surplus value of Hawala bankers is not restricted to providing financial services, but that many of them are able to smooth out potential difficulties in the settlement of illegal transactions. Hans Nelen and Francien Lankhorst highlight the dilemmas facing the legal and financial professions in their professional relationships with criminal clients. The primary focus of their contribution is on the position and the role of lawyers and notaries public.

The last part of the book is dedicated to the containment of organized crime. The first two contributions in this part are all related to direct policies which are anchored in criminal and criminal procedural law and primarily implemented by the criminal justice system. Richard Staring discusses the process of criminalization of human smuggling in the Netherlands in the context of a number of national and international developments. Carlo Morselli, Dave Tanguay, and Anne-Marie Labalette look into criminal conflicts, collective violence, and how to deal with outbursts of violent confrontations. Their analysis of violent conflicts between biker gangs in the province of Quebec indicates that increases in account settlements are due to cyclical transitions from individual to group-based violence in dispute settlements. They argue that external intervention from specialized law-enforcement squads was required to bring the persisting conflict to a close.

As Fijnaut and Paoli (2004) already showed in *Organised Crime in Europe: Concepts, Patterns and Policies in the European Union and Beyond*, significant differences between legal and administrative systems and regulations in the various countries make a uniform policy towards organized crime difficult, if not impossible. Nevertheless, internationally, a development can be observed in the direction of a *dual strategy*, i.e. a strategy in which law enforcement powers and administrative powers complement one another.

The dual strategy originated in New York City. The last three chapters are dedicated to various manifestations of such a strategy. In his contribution on labour racketeering in the United States, Jim Jacobs argues that the civil RICO-law offered many opportunities to break the power that the Cosa Nostra organized crime families exerted over many American labour unions for most of the twentieth century. In Italy and the Netherlands, more emphasis has been put during the last decade on an administrative approach designed to prevent the facilitation of organized crime. This approach consists of a number of instruments, ranging from integrity testing of civil servants, the refusal or withdrawal of permits, to screening and exclusion of companies that compete for major contracts. Antonio la Spina analyses indirect anti-mafia strategies that were introduced in Italy throughout the last two decades. Hans Nelen and Wim Huisman describe why the administrative approach in Amsterdam is seen as an important addition to more traditional ways of combating organized crime, but they warn against unrealistic expectations. As well as La Spina in his contribution on the Italian case, Nelen and Huisman emphasise that civil and administrative instruments and policies should not be regarded as 'alternatives' to large scale penal-action, but rather as complementary or integrated approaches to organized crime.

Criminal Groups and Activities

"The world is yours, select your country, Tony" (Sopranos)

Historical Transformations

Chapter 1
Reflections on the Sicilian Mafia: Peripheries and Their Impact on Centres

Anton Blok

The Setting

The term "mafia" comes originally from Sicily, where it refers to the private use of violence in public domains. Its etymology is controversial (Novacco, 1959). Whatever its feudal and patrimonial dispositions, mafia is a modern phenomenon: it developed in the slipstream of Italian unification when a modern state structure imposed itself on a predominantly agrarian society still largely feudal in its basic features. In the absence of effective central control over the means of violence, mafia took shape as an *instrumentum regni* of Italian politicians, who chose to rule Sicily through its dominant class of absentee landlords most of whom were residing in Palermo, long the centre of wealth and power of the island. (Incidentally, this may help explain why eastern Sicily always had a much lower density of mafia than the four western provinces, which formed the hinterland of the island's capital.)

Mafia presence was particularly strong in and around Palermo, in the adjacent citrus fruit area of the Conca d'Oro (Golden Shell) and its vast forbidding hinterland that covered the island's western interior, aptly characterized by Fentress (2000: 132) as a "bandit corridor" because of the prevailing cattle theft "industry," organized through a vast network of accomplices that extended from the western part of Palermo province through Caltanissetta province and into Agrigento province (Fentress, 2000: 132). This was the area of large cereal-pasture holdings (the so-called latifundia or ex-feudi), the history of which goes back to early antiquity. Until the 1960s, absentee landowners entrusted its management to ruthless local upwardly mobile leaseholders and overseers who had a reputation for violence and therefore enjoyed "respect." These local bosses and their retainers formed the backbone of the rural mafia, which consisted of loosely formed coalitions, called *cosche* or "families." Each family or *cosca* controlled its own territory – a village, a town, or an urban neighbourhood – and enforced a tribute on all economic transitions under its "protection." The typical mafioso was a middleman or power broker, who thrived on the gaps in communication between landlord and peasant, seller and buyer of commodities and services, political candidate and electorate. Ultimately, mafiosi held sway over the links between state and community, backing up their protection and mediation by violence.

D. Siegel and H. Nelen (eds.), *Organized Crime: Culture, Markets and Policies.*
© Springer 2008

Profile

Mafiosi controlled local resources, like property, markets, services, and votes. They operated in collusion with members of the elite, most notably urban-based landlords, politicians, government officials, and businessmen. This "contiguity" involved mutual protection and the exchange of "favours." But the protection which mafiosi offered was sometimes difficult to distinguish from extortion and the boundaries between victim and accomplice were often likewise blurred. In the no mans land between public and private domains, mafiosi were left a free hand in the management of local affairs. Although they exacerbated class tensions between landlords and peasants by their rent-capitalist management of the estates and appropriating land and becoming landowners in their own right, they also controlled these tensions by carving out channels of upward mobility for ambitious and ruthless peasant and shepherds. In Sicily, as probably elsewhere under similar conditions (Brogan, 1998; Tilly, 1997, 2000; Varese, 2001), mafia and politics provided "carrières ouvertes aux talents" [careers open to those with talents] (Fentress, 2000: 149). Towards the large mass of landless peasants and shepherds, from whose ranks they usually originated, their attitude involved undisguised disdain and exploitation. When indicted for violent crimes, mafiosi were usually acquitted for lack of evidence because of high-level protection and because local witnesses would rarely openly turn against them. This greatly helped to enhance their power and their reputation as "men of respect." Inspired by both fear and admiration, the local population drew up a "wall of silence" (omertà), which ultimately blocked effective prosecution of mafiosi. Until recently, the power of mafiosi, although surrounded and buttressed by silence, was openly displayed. It illustrated the peaceful coexistence between mafia and state.

Far from being "a state within a state," as magistrates and journalists often represented the phenomenon, mafiosi successfully infiltrated public institutions, including political parties, local governments, the judiciary, banks, and legal firms. They did so through their own personal network of "friends" rather than as members of corporate groups or of a unified and centralized organization. With the extension of the suffrage, they enlarged their grip on the electorate and controlled more votes. With the extension of postwar government aid to "develop" the country's southern peripheries, mafiosi swallowed up ever more funds, most notably in the urban construction industry, always capable of placing themselves with cunning and force between state and citizen (cf. Jamieson, 2000: 20–21). Sicilian mafia appears as the violent alternative to civil society. This was the price the Italian State eventually had to pay for the pragmatic accommodation, which later became know as "pollution" of public institutions. It was the other side of a peaceful coexistence. It also provides an example of the great impact a periphery can have on the centre. Long played down, dismissed, or wilfully misunderstood as a construction – as an invention of people who do not like Sicily – mafia was only recently recognized as one of Italy's main social problems. Yet this reference to the "the Southern Problem" (or "Question") still ignores and obscures the extent to which the Italian state itself has been transformed in its interaction with its southern periphery.

Representations: Hierarchy Versus Segmentation

The persistent and popular representations of mafia as "a state within a state" and as a centralized monolithic organization are wide off the mark. They make too much of the organization and too less at the same time. Presenting mafia as a single unified structure neglects its structural flexibility and fluidity manifest in open-ended networks and action sets, as well as its parasitic relationship with the state. To understand how mafia works is to start the investigation at the local level, because what the phenomenon in Sicily comes down to are local cliques structured by ties of kinship, marriage, and friendship, who control local resources with violent methods while enjoying large measures of impunity because of their contiguity with powerful protectors at higher levels of society who need local strongmen as managers of their properties and as canvassers of votes (Blok, 1974). The relationships between these "families" are characterized by conflict and accommodation – strikingly similar to the relationship between states – rather than supervised, coordinated, or controlled from the top by a commission of sorts, as current expressions like "*Cosa Nostra*" (Our Thing) and "The Mafia" suggest. In his respect the alliance of mafia families in the United States (largely from Sicilian extraction), from whom the Sicilians adopted this denomination, shows much greater stability (Janni, 1972; Sterling, 1990).

The idea that the *cosche* operated like sovereignties is attested by the failed attempts to coordinate their relations from above. When the so-called "Commission" of *Cosa Nostra* was put together to contain intense intra-mafia violence and to impose a *pax mafiosa*, as happened for the first time in the late 1950s and again on several occasions in the 1970s, it fell apart each time because representatives of the various factions could not agree on an overall policy, or tried to outmanoeuvre each other and dominate the Commission. In June 1963, what had remained of this board blew up in the notorious Ciaculli affair that brought an end to the first mafia war and heralded an era of anti-mafia policy (Dickie, 2004: 305ff; Jamieson, 2000: 16ff; Sterling, 1990: 101ff). The second mafia war unfolded around 1980 and ended in the near extermination of one of the warring factions – leaving about thousand mafiosi dead in 2 years. These failings alone clearly demonstrate that the Sicilian mafia cannot be understood as a single unitary organization (cf. Fentress, 2000: 246). An analysis of these episodes suggests the image of a changing configuration of alliances in and between local "families." In all these instances, incipient hierarchy gave way to segmentation (cf. Catanzaro, 1992: 198–203; Paoli, 2003: 62–64).

In a perceptive essay, Moss (2001: 325) draws attention to the differences in the accounts of one and the same *pentito* depending on whom he is talking to: "Arlacchi notes how greatly the account given to magistrates by the mafia *pentito* Antonino Calderone, which stresses the accepted organizational hierarchy, division of labor and obedience to rules, differed from the account which Calderone provided to Arlacchi himself, which was full of the coincidences, breaches of conventions, betrayals and few-holds-barred struggles of actual mafia behaviour. Emphasis on the casual and coincidental rather than the logic of organizational hierarchy tends of course to undermine the broad perspective adopted in the legal version…" (Moss, 2001; cf. Arlacchi, 1992: v–vi). The interviews Bettini (1994) had with the young mafioso Enzo from Castelvetrano confirm mafia practices highlighted in Arlacchi's interviews with Calderone. Networking and building action sets take priority over allegiance to corporate groups: "The rules of Cosa Nostra, the links with chiefs of 'families' count for nothing with respect to the supreme law of the mafia: that of the strongest" (Bettini, 1994: 270–271). Drawing on multiple confessions of pentiti in the 1980s and 1990s, the recent literature on mafia makes much of swearing the oath of loyalty in initiation rituals that mark the passage into local "families." Yet considering the intense rate of intra-mafia betrayal and violence, both between and within "families," these rituals testify to a justified lack of mutual trust between mafiosi rather than to its supposed effect of loyalty. As pragmatic insider Tommaso Buscetta puts it, "The man who stands besides you might take you to your tomb as easily as he would take you to a party" (Arlacchi, 1994: 155; cf. Bettini, 1994: 73ff; Paoli, 2003: 118–120; Stille, 1995: 99–120, 303ff).

Transitions

Over the past three decades, the mafia in Sicily gradually lost its role as a pragmatic extension of the state and assumed the character of a hidden power. Mafiosi disappeared as public figures with public identities. The mafia's peaceful coexistence with the state came to an end. Terror replaced accommodation. Like outlaws elsewhere, mafiosi have to hide from the law (if they did not, as pentiti, outright cooperate with the judicial authorities under a protection program), living clandestine lives in unassuming locations, including isolated farms. The Corleonesi Salvatore Riina and his colleague Bernardo Provenzano are the most famous examples of such outlaws: not only because they were major leaders, but also because it took several decades before they could be arrested. If anything, this betrays long periods of effective protection, both from inside and outside the mafia – if one can, in this case, tell one from the other. Several circumstances are responsible for this major transition.

First, beginning in the early 1970s, Sicilian mafiosi moved into the international trade in narcotics which produced huge and fast profits but entailed high risks. These ventures, like the building boom in the early 1960s, resulted in internecine struggles between rival factions – between upstarts from the Palermo hinterland, most notably the Corleonesi led by Luciano Leggio and Salvatore Riina, and the urban establishment in Palermo of which Stefano Bontate was one of the most striking figures, next to his posh and wealthy friends the cousins and tax-collectors Salvo. State interference provoked campaigns of terror against state representatives, including politicians, magistrates, and policemen (Jamieson, 2000: 27). This new development brought mafiosi into open conflict with the Italian state, which could not stand back and accept open warfare on its territory and close its eyes for the massive infiltration of drug money in its economy. Nor could the Italian government ignore the pressure from neighbouring states and overseas countries to take a strong position against the drug traffic and the laundering of money.

The extent to which the tables had been turned on politicians who had either directly or indirectly protected the interests of mafiosi who provided them with electoral support, can be illustrated by the following anecdote. After the first Sicilian politicians and magistrates who tried to withdraw from the mafia or to combat it with determination, were assassinated in public in 1979 and 1980 to be remembered as the first wave of "excellent cadavers," Stefano Bontade, the flamboyant mafia leader in Palermo allegedly warned the then prime minister Giulio Andreotti on his visit to the island in February 1981 with the following words: "We command in Sicily and if you don't want the DC to disappear completely you'll have to do as we say. Otherwise we'll take not only your votes in Sicily but the votes in Reggio Calabria and the whole of the south of Italy as well. You'll have to make do with the votes in the north where everyone votes Communist" (Jamieson, 2000: 222; cf. Stille, 1995).

Yet helped by a new generation of magistrates and special legislation (which included the coinage of the term *associazione mafiosa*), the Italian state, from the early 1980s onwards, started its crackdown on the mafia, which culminated in effective proceedings (the so-called maxi-trials) against hundreds of mafiosi. The prosecution was substantially helped by mafiosi from losing factions who, partly for reasons of revenge and partly in order to survive, repented (hence their name *"pentiti"*), turned state's evidence, and helped build a case against their former peers (Moss, 2001; Schneider & Schneider, 2003; Stille, 1995).

Behind all these changes lurked an even larger transition. This was the demise of communism in Europe, which unintentionally undermined the mafia's raison d'etre. From the immediate postwar years onwards, mafiosi had always supported political parties that took a strong position against communist and socialist parties. Faced with the largest communist party in Western Europe, the Italian Christian Democrats could not forego the support of an informal power that provided huge blocks of votes and thus effectively staved off what was perceived as "the danger of communism" in a predominantly agrarian and Roman Catholic region. With the disintegration of the former Soviet Union and the end of the Cold War, the situation radically changed and the shield of protection that kept mafiosi out of prison began to fall apart. The situation resembled the crackdown on the mafia under Fascism. Prefect Mori's large-scale operation in the late 1920s took place after the Sicilian landed classes had found more effective and less costly protection of their property from the new (totalitarian) government. This turn exposed mafiosi and left them without protection (Duggan, 1989; cf. Petacco, 1972: 18–26; Servadio, 1976: 57–63).

One cannot help notice another parallel in the different ways the Italian state proceeded against the mafia. In both cases, prosecutors represented the mafia as a corporate organization, as a single, unified, hierarchical organization, controlled from the top by one or more super chiefs (so-called '*capi di tutti capi*), who coordinated the relations between local "families." As noted above, there is little evidence for this so-called "Buscetta theorem".[1] Although recurring tendencies towards centralization were not lacking, their very failure shows the strength and structural continuity of local groups tied together by links of kinship, marriage, and friendship. These *cosche* or "families" were embedded in larger networks from which crystallized changing coalitions and specific action sets. Moreover, these larger networks of which mafiosi were part also included people from outside their ranks: patrons, friends, and clients from any social station, all willing to exchange "favours." Without taking into consideration this collusion, also called "contiguity" or "cohabitation," with the wider world, it would be difficult to develop a realistic perception of mafia. Major crimes involved the organization of specific and unique action sets, of which not always all participants knew each other, or even had an idea about the larger goal of the operation. A telling example provides Tommaso Buscetta when describing the steps that led to the elimination of Enrico Mattei in October 1962 (Arlacchi, 1994: 80–83).

The very absence of a unified and centralized structure may have considerably served the mafia and help account for its remarkable longevity in modern society (cf. Blok, 1974; Catanzaro, 1992). Segmentation rather than hierarchy reflects the same paucity of communications that made it possible for mafiosi to place themselves as power brokers at decisive junctions in the relationships between local and national levels, which in turn hampered further integration within the framework of the Italian state. More recently, it has been convincingly argued that "despite the experiments [with mafia centralization] carried out in the last four decades of the twentieth century, the ruling bodies of single families remain the real centers of mafia power, and segmentation is the prevalent form of differentiation" (Paoli, 2003: 64).

Conclusions

Today, the term "mafia" is widely used as a synonym for organized crime. A still larger conflation of the term includes (ironic) references to any coterie wielding illicit power in formal organizations and institutions, including governments, business firms, and universities. To contain the growing inflation of the terminology, one may restrict the use of the term "mafia" to denote a form of organized crime that includes collusion and contiguity with persons who represent public institutions. If we agree that mafiosi are agents who operate with violent methods in the border

---

[1] The Dutch professor of criminal law at the University of Utrecht Geert-Jan Knoops, emphasized the problems attending the prosecution of delinquents who operate in groups (action sets) that take a different form in each situation, such as the *Hells Angels*. In this case it is impossible to prove the existence of a durable structured organization aimed at committing crimes. Criminal justice cannot but resort to contestable constructions to reach a verdict (NRC 18 October 2005). Presenting the Sicilian mafia as a unified and centralized organization comes close to such a construction. In the recent trial against the Dutch *Hells Angels* the court abstained from this approach, a decision that remained controversial.

zones of public and private domains to enrich themselves and convert their wealth into honour and social status, there is no need to widen further the meaning of the term to include any form of organized crime, because the private use of violence in public places has lately expanded on a world scale in Latin America, Eastern Europe, the former Soviet Union, Southeast Asia, and various African countries. All these cases are unique, but one cannot help noticing their family resemblances, including the transformative impact peripheries may have on their centres.

References

Arlacchi, P. (1992). *Gli uomini del disonore. La mafia siciliana nella vita del grande pentito Antonino Calderone*. Milano: Mondadori.

Arlacchi, P. (1994). *Addio Cosa Nostra. La vita di Tommaso Buscetta*. Milano: Rizzoto.

Bettini, M. (1994). *Pentito. Una storia di mafia*. Torino: Bollati Boringhieri.

Blok, A. (1974). *The mafia of a Sicilian village, 1860–1960. A study of violent peasant entrepreneurs*. Oxford: Basil Blackwell.

Brogan, P. (1998). The drugs war (Colombia). In P. Brogan (Ed.), *World conflicts*. London: Bloomsbury.

Catanzaro, R. (1992). *Men of respect. A social history of the Sicilian mafia*. New York: Free Press.

Dickie, J. (2004). *Cosa Nostra. A history of the Sicilian Mafia*. London: Houghton & Stoughton.

Duggan, C. (1989). *Fascism and the mafia*. New Haven: Yale University Press.

Fentress, J. (2000). *Rebels and mafiosi. Death in the Sicilian landscape*. Ithaca: Cornell University Press.

Janni, F. (1972). *A family business. Kinship and social control in organized crime*. New York: Russell Sage.

Jamieson, A. (2000). *The Anti-mafia. Italy's fight against organized crime*. London: Macmillan.

Moss, D. (2001). The gift of repentance: A Maussian perspective on twenty years of pentimento in Italy. *Archives européennes de sociologie, 42*(2), 297–331.

Novacco, D. (1959). Considerazioni sulla fortuna del termine 'mafia'. *Belfagor, 14*, 2–11.

Paoli, L. (2003). *Mafia brotherhoods. Organized crime Italian style*. Oxford: Oxford University Press.

Petacco, A. (1972). *Joe Petrosino*. Milano: Mondadori.

Schneider, J., & Schneider, P. (2003). *Reversible destiny. Mafia, antimafia, and the struggle for Palermo*. Berkeley: University of California Press.

Servadio, G. (1976). *Mafioso. A history of the mafia from its origins to the present day*. London: Secker & Warburg.

Sterling, C. (1990). *Octopus. The long reach of the international mafia*. New York: Simon & Schuster.

Stille, A. (1995). *Excellent cadavers. The mafia and the death of the first Italian Republic*. London: Jonathan Cape.

Tilly, C. (1997). War making and state making as organized crime. In C. Tilly (Ed.), *Roads from past to future* (pp. 165–191). Lanham, MD: Rowman & Littlefield.

Tilly, C. (2000). Preface to the second edition of Anton Blok, La mafia di un villaggio siciliano, 1860–1960. Torino: Einaudi/Comunità.

Varese, F. (2001). Is Sicily the future of Russia? Private protection and the rise of the Russian Mafia. *Archives européennes de sociologie, 42* (1), 186–220.

Chapter 2
The Decline of the Italian Mafia

Letizia Paoli

Introduction

In the Italian public debate, Italian mafia organisations, and particularly the Sicilian Cosa Nostra, are still routinely portrayed as immensely powerful and almost invincible. The few observers who express less pessimistic views are harshly criticised by antimafia prosecutors and activists alike and accused of having lost contact with local southern Italian reality or, even worse, of serving mafia interests. At the same time, with varying degrees of good faith, most members of the law enforcement community, antimafia NGOs and academic and journalistic commentators insist upon the fact that after an unprecedented peak in the early 1990s antimafia fight has dramatically slowed down and weakened (cf. Bolzoni & Lodato, 1998; Lodato & Grasso, 2001).

Breaking a taboo, this contribution aims to document and explain the current crisis undergone by the two largest, longest-standing and most powerful forms of Italian organised crime, the Sicilian Cosa Nostra and the Calabrian 'Ndrangheta. As shown in the following pages, mafia groups' current decline is primarily the result of the intensification of law enforcement repression since the early 1990s. Contrary to popular perceptions, my analysis shows that *direct* antimafia policies,[1] which are anchored in

---

[1] Antimafia policies can be distinguished in direct and indirect, a distinction originally advanced by La Spina (2004). While the direct policies primarily consist in law-enforcement repression and sanctioning, the indirect ones address civil society and the public administration, ranging from anti-racket legislation to the dissolution of local councils infiltrated by the mafia and to initiatives aimed at spreading a "legality culture" and promote security in southern Italy. As shown elsewhere (La Spina, 2004, this volume; Paoli, 2007), indirect antimafia policies have been less effective than direct ones. Here it is also important to stress that the distinction between direct and indirect antimafia policies is analytic, because in practice there are many overlappings between them.

criminal and criminal procedural law and primarily implemented by the criminal justice system, have continued to be quite effective at the beginning of the twenty-first century. As we will see in the following pages, such a conclusion is based on the analysis of a plurality of quantitative and qualitative data, including the number of *mafiosi* arrested and convicted, the amount of assets seized and confiscated, mafia-related murders and qualitative evidence resulting from law enforcement investigations, mafia defectors' statements and short letters (the so-called *pizzini*, see infra) written by high-ranking bosses of the Sicilian Cosa Nostra.

Though accelerated by law enforcement action, mafia organisations' decline has also been fostered by contradictions and shortcomings in their own legitimation system. Cultural norms have in fact braked mafia groups' expansion outside of their home regions and the diversification of their illicit activities. As a result of both law enforcement and culture, southern Italian mafia organisations, with the partial exception of some 'Ndrangheta families, are now excluded from large-scale international flows of illegal trade and have receded into their own territories, focusing instead on racketeering in their local communities. Particularly in northern and central Italy, they have lost terrain to a myriad of new criminal enterprises that are much more ephemeral and often composed of foreign migrants (cf. Paoli, 2004, 2007 for an analysis of Italy's non-mafia forms of organised crime).

In the first section I discuss the main characteristics of the two largest and most powerful mafia organisations; the Sicilian Cosa Nostra and the Calabrian 'Ndrangheta, emphasizing their peculiarities vis-à-vis other forms of European organised crime. In the latter part of that section, I also show how cultural codes effectively restrain mafia groups' economic activities. In the second section I briefly recall the intensification of the antimafia fight since 1992 and then examine how these two organisations have reacted to it. Some concluding remarks follow.

Cosa Nostra and the 'Ndrangheta: What Makes Them Different from Most Other Forms of Organised Crime?

Contrary to what most scholars maintained up to the early 1980s, judicial inquiries carried out since then have demonstrated that formalised mafia groups do exist. Cosa Nostra (Our Thing) and the 'Ndrangheta (Society of the Men of Honour) are the two largest and most stable criminal organisations and are each composed of about a hundred mafia groups.[2] Though it is not possible to establish clear lines

---

[2] Cosa Nostra and the'Ndrangheta do not exhaust the panorama of organised crime even in their home regions. In Sicily and northern Calabria several other mafia and pseudo-mafia groups are active and in Sicily some of them have even successfully challenged the local Cosa Nostra families. In addition to the Sicilian and Calabrian groups, two other clusters of crime groups are usually referred to as organised crime in Italy: (1) the "galaxy" of mafia-like and gangster-like groupings in Campania, collectively known as "camorra" and (2) the multiplicity of criminal groups, gangs and white-collar criminal networks operating in Apulia. Other non-traditional organised crime groups and networks, partially composed by foreign migrants, are also active since the early 1990s, primarily in the Centre-North (for more information, see Paoli, 2004, 2007).

of continuity, historical research since the late 1980s has demonstrated that antecedents of the contemporary mafia groups existed in the 1880s, if not before (cf. Pezzino, 1987). In the mid-1990s the members of Cosa Nostra and the 'Ndrangheta were estimated at about 3,500 and 5,000 males, respectively (Paoli, 2003: 30–32). No newer estimates have been published since then.

Secret Brotherhoods

Cosa Nostra and the 'Ndrangheta possess the distinguishing trait of organisations (Weber [1922] 1978: 48): independent government bodies that regulate the internal life of each associated family and that are clearly different from the authority structure of their members' biological families. Starting from the 1950s, moreover, super-ordinate bodies of coordination were set up – first in the Cosa Nostra, then, since the early 1990s, in the 'Ndrangheta as well. Composed of the most important family chiefs, they are known as "commissions." Despite the media and judicial emphasis on them, the powers of these collegial bodies have always been quite limited, as their primary and almost exclusive task has been the regulation of violence against mafia members and public officials.

Neither the Cosa Nostra nor the 'Ndrangheta can be assimilated to Max Weber's ideal type of legal-rational bureaucracy, as was suggested by Donald Cressey (1969) in the late 1960s with reference to the American La Cosa Nostra. Far from recruiting their staff and organising the latter's work according to the criteria and procedures of modern bureaucracies, mafia groups impose a veritable "status contract" on their members (Weber, 1978: 672). With the ritual initiation into a mafia group, the novice is required to assume a new identity permanently – to become a "man of honour" – and to subordinate all his previous allegiances to the mafia membership. If necessary, he must be ready to sacrifice even his life for the mafia family.

The "men of honour" in Sicily and Calabria are obliged to keep secret the composition, the action, and the strategies of their mafia group. In Cosa Nostra, in particular, the duty of silence is absolute. Secrecy constitutes, above all, a defence strategy. Since the unification of Italy in 1861 mafia groups have been at least formally criminalised by the state and, in order to protect themselves from arrest and criminal prosecution for their continuing recourse to violence, they have needed to resort to varying degrees of secrecy.

The ceremony of affiliation additionally creates ritual ties of brotherhood among the members of a mafia family: the "status contract" is also an act of fraternisation (Weber, 1978: 672). The new recruits become "brothers" to all members and share what anthropologists call a "regime of generalised reciprocity": this presupposes altruistic behaviour without expecting any short-term reward. As Lestingi (1884), chief prosecutor for the monarchy, pointed out in 1884, mafia groups constitute brotherhoods whose "essential character" lies in "mutual aid without limits and without measure, and even in crimes" (Lestingi, 1884: 453).

As secret brotherhoods using violence, southern Italian mafia confederations have remarkable similarities to organisations such as the Chinese Triads and the

Japanese Yakuza. With their centuries-old histories, articulated structures, and sophisticated ritual and symbolic apparatuses, all these organisations – and the American descendant of the Sicilian Cosa Nostra – have few parallels in the world of organised crime. None of the other groups that systematically traffic in illegal commodities have the same degree of complexity and longevity (Paoli, 2002b).

The Will to Power

Cosa Nostra and the 'Ndrangheta share another important peculiarity with the Chinese Triads and the Japanese Yakuza. Unlike other contemporary organised crime groups, they do not content themselves with producing and selling illegal goods and services. Though these activities have acquired an increasing relevance over the past 30 years, neither the trade in illegal commodities nor the maximisation of profits has ever been the primary goal of these organisations. As a matter of fact, at least in the case of southern Italian mafia coalitions, it is hardly possible to identify a single goal. The Sicilian Cosa Nostra and the Calabrian 'Ndrangheta are multi-functional organisations. In the past 100 years, their members have exploited the strength of mafia bonds to pursue various endeavours and to accomplish the most disparate tasks. Already in 1876 the Tuscan aristocrat Leopoldo Franchetti ([1876] 1993: 100) pointed out the "extraordinary elasticity" of the Sicilian associations of *malfattori* (evildoers): "the goals multiply, the field of action widens, without the need to multiply the statutes; the association divides for certain goals, remains united for others."

Among these tasks the exercise of political domination has always been pre-eminent. The ruling bodies of Cosa Nostra and the 'Ndrangheta claim, above all, an absolute power over their members. They control every aspect of their members' lives, and they aim to exercise a similar power over the communities where their members reside. For a long time their power had a higher degree of effectiveness and legitimacy than that exercised by the state. In western Sicily and in southern Calabria mafia groups successfully policed the general population, settling conflicts, recovering stolen goods, and enforcing property rights.

Even today, although most mafia rules are no longer systematically enforced, mafia families exercise a certain "sovereignty" through a generalised system of extortion. As a state would do, they tax the main productive activities carried out within their territory (Paoli, 2003: 154–172). Moreover, whenever *mafiosi* are asked to mediate conflicts, guarantee property rights and enforce rules compatible with their own legal order, they do not hesitate to intervene. Even contemporary "men of honour" still take these duties seriously (cf. Lodato, 1999: 73).

The political dimension of mafia power is also proven by the fact that in the second half of the twentieth century southern Italy's mafia organisations have participated in at least three plots organised by right-wing terrorist groups. Moreover, since the late 1970s, Cosa Nostra has assassinated dozens of policemen, magistrates, and politicians. The mafia challenge to state power reached a climax in the early 1990s. In 1992, Cosa Nostra murdered the Palermitan judges Giovanni Falcone and Paolo Borsellino in two spectacular bomb explosions. In 1993, in an effort to demonstrate the national power of the mafia, a series of terrorist bombings occurred – for the first time out of traditional mafia strongholds – in Rome, Florence, and Milan.

The Incomplete Entrepreneurial Transformation

Despite their power, Cosa Nostra and the 'Ndrangheta have been unable to guarantee themselves a monopoly in any sector of the illegal economy outside southern Italy. In the early 1980s, Cosa Nostra families played a pivotal role in the transcontinental heroin trade from Asia to the United States via Sicily. But in the second half of that decade, they lost this position after being targeted by law-enforcement investigations and replaced in the U.S. market by a plethora of Mexican, Chinese, and, more recently, Colombian heroin suppliers (Paoli, 2003: 215–216).

The power of Cosa Nostra is not unchallenged even within its strongholds. Given the extreme rigidity of their recruitment policies, in fact, Cosa Nostra families often find themselves in a minority position and are hence unable to control the whole underworld. This difficulty was admitted to even by Giovanni Brusca, the man who was supposed to become Totò Riina's successor in Cosa Nostra leadership, but who in fact became a mafia witness after his arrest in 1994:

> Many believe that Cosa Nostra heads all criminal activities. That in Palermo or in Sicily every illegal activity is controlled by the *mafiosi*. People believe that prostitution and bur- glaries, bank robberies, and car thefts are all entries in the budget of the Mafia Inc. Those that I have just listed are external activities, known about, tolerated, and controlled by men of honour. But they are separate worlds, which only rarely come into contact with each other. In some cases, there might be some collaboration, but this is only in very special cases (Lodato, 1999: 67).

Despite the growing relevance of economic activities, "the mafia has not become a set of criminal *enterprises*" (Becchi & Turvani, 1993: 156). Its history as well as its cultural and normative apparatus prevent this transformation and today constitute a constraint as much as a resource. By building a strong collective identity, shared cultural codes and norms enhance group cohesion and create trustful relationships among mafia members. The reliance on status and fraternisation contracts, which are non-specific and long-term, produces a high degree of flexibility and makes the multi-functionality of mafia groups possible. The same shared cultural codes and norms also represent a powerful brake on entrepreneurial initiative. The prohibition on exploiting prostitution, for example, which exists in both confederations, has blocked the entrance of the Cosa Nostra and 'Ndrangheta groups into what has become one of the most profitable illicit trades: the smuggling of humans and the exploitation of migrants in the sex industry.

Especially constraining is one of the preconditions for recruitment: only men born either in Sicily or in Calabria or descending from mafia families can be admitted as members. This rule has long prevented Cosa Nostra and 'Ndrangheta families from adding new members with the experience necessary to compete in the black markets for arms, money, and gold. Rigid recruitment criteria have also hampered the geographical expansion of mafia power. Cosa Nostra, for example, prohibits settling families outside of Sicily and the only partial exception to this rule has been represented by the Cuntrera-Caruana family from Siculiana. This self-imposed rule, which aims to strengthen the cohesion of the mafia consortium, has limited its involvement in the international narcotics trade – currently the largest of the illegal markets. 'Ndrangheta families, thanks to their extensive branches in northern Italy and abroad, played a larger role in narcotics trafficking in the 1990s, importing large quantities of cocaine and hashish from Latin America and North Africa. Today, however, the 'Ndrangheta faces new competition from foreign and Italian traffickers with more direct connections to drug-producing and transit countries (Paoli, 2003: 217).

The "will to power" of the mafia organisations also negatively affects security and business decisions, as a leading Palermitan prosecutor pointed out in 1992:

The true goal is power. The obscure evil of organisation chiefs is not the thirst for money, but the thirst for power. The most important fugitives could enjoy a luxurious life abroad until the end of their days. Instead they remain in Palermo, hunted, in danger of being caught or being killed by internal dissidents, in order to prevent the loss of their territorial control and not run the risk of being deposed (Scarpinato, 1992: 45).

As a result, since the early 1990s Cosa Nostra and 'Ndrangheta families have extracted a growing percentage of their income from entrepreneurial activities that depend on the exercise of regional political domination. They practice systematic extortion in their communities and, thanks to intimidation and collusion with corrupt politicians, they have struggled to control the market for public works. Their interest originally focused on the construction market: from the 1950s onwards, Cosa Nostra groups, and later on 'Ndrangheta families as well, set up their own building companies and pressed claims to become directly involved in both small and large building sites. Since the 1980s, moreover, mafia enterprises of both regions have become eager to obtain a share of any public work tendered by national or local public administrations.

In addition to the spread of mafia interests, the novelty of the 1980s and early 1990s was represented by the involvement of mafia representatives in the *comitati d'affari*, originally formed from politicians and entrepreneurs, which had controlled the bidding processes of large-scale public works all over the country for many years. Thus mafia influence no longer took place only "downstream," i.e., at the end of the economic process of public investment (subcontracts and extortions). It was also exercised "upstream" at the beginning of the process, with decisions made jointly by mafia representatives, state agencies and the large building companies that were particularly interested in obtaining large contracts for public works. In western Sicily, for example, a sort of "duopoly" was established in the late 1980s and early 1990s: the public works market was subject to the complete "top-down" control of two strong groups – Cosa Nostra and the *comitati d'affari* – which had joined forces in a kind of symbiosis cemented by silence and complicity (Paoli, 2003: 147–148, 174–175).

Unlike other western forms of organised crime, the meaning (and danger) of the two main Sicilian and Calabrian mafia organisations cannot be limited to their involvement in illegal markets. Their peculiarity lies in their will to exercise political power and their interest in exercising sovereign control over the people in their communities.

Law-Enforcement Repression and Mafia Organisations' Reactions

Cosa Nostra and the 'Ndrangheta's economic activities, political relationships and even their overall associational life have been seriously hit by the intensification of law enforcement action that followed the "terrorist" attacks staged by Cosa Nostra in 1992–1993. The attacks against Cosa Nostra were particularly devastating, as allegedly recognised even by Bernardo Provenzano, the latest known "*capo*," or chief of the Sicilian mafia organisation, who was captured in April 2006. According to an informant, in fact, in the second half of the 1990s, Provenzano said that he was convinced that Cosa Nostra would need at least 5–7 years to recover from the serious crisis into which it had plunged and to improve its economic situation, which was at that point precarious (Ministero dell'Interno, 2001b: 10).

The State Counterattacks After 2002

Provenzano's arrest in 2006 crowned the intensified law enforcement efforts to capture all the leading Cosa Nostra figures of the 1980s and 1990s, some of whom had been on the run for decades. The first outcome of such an effort and a major boost for the whole antimafia campaign had been the arrest in January 2003 of Totò Riina, Provenzano's predecessor at the head of Cosa Nostra, after more than 22 years spent in hiding. Between these two spectacular arrests, hundreds of less-known mafia members were also arrested and tried. The Direzione Investigativa Antimafia (DIA, 2007) reports that in the period 1992–June 2006 it allowed prosecutor's offices to issue 1,627 arrest warrants against members of Sicilian mafia groups and 2,317 against members of the Calabrian mafia. These data do not of course represent the total number of arrested persons. Another partial source is the Ministry of the Interior, which reports the data concerning the dangerous fugitives arrested, most of whom are mafia bosses or at least mafia members. As Table 2.1 shows, between 1992 and June 2005 over 1,200 fugitives were arrested, who belonged to either Sicilian or Calabrian mafia groups.

Neither the DIA nor the Ministry of the Interior distinguish between members of Cosa Nostra and 'Ndrangheta families and members of other mafia or pseudo-mafia

Table 2.1 Arrested fugitives belonging to Calabrian or Sicilian mafia groups – 1992–2005

	Calabrian mafia groups	Sicilian mafia groups
1992	29	12
1993	47	69
1994	34	71
1995	59	90
1996	66	56
1997	34	39
1998	86	80
1999	53	50
2000	43	38
2001	44	25
2002	38	25
2003	22	26
2004	30	18
2005 (6 months)	16	5
Total	601	604

Source: For 1992–1997, Ministero dell'Interno (2001a: 438); for 1997–2005, Ministero dell'Interno (2005: 26–27)

crime groups active in Sicily. However, there are no doubts that the former account for the bulk of mafia-type organised crime active in the two regions and they are prioritised by law enforcement agencies. Given the previous estimates of 3,500 and 5,000 males respectively composing Cosa Nostra and the 'Ndrangheta, it is clear that the probability of being arrested has become very high.

Members of Cosa Nostra and 'Ndrangheta are not only arrested but also charged and convicted with heavy sentences. Even in Calabria, where state repression has until recently not received much support from civil society, mafia groups have experienced severe restrictions. In January 1999, at the end of the hearings concerning the Olimpia-1 Operation, the Reggio Calabria court handed down 62 life sentences and 141 sentences amounting to over 1,380 years of imprisonment, while another three hundred defendants are involved in the three following parts of the inquiry. Likewise, the 99 defendants of the Tirreno maxi-trial, which took place in Palmi against the greatly feared Piromalli and Molè families, were sentenced to 89 life sentences and 731 years of imprisonment by the local first-degree court. The investigations did not focus only on the core families of the province of Reggio Calabria, but also involved their links in central and northern Italy. In Milan, for example, between 1994 and 1998 more than a thousand members of the 'Ndrangheta faced trial in about twenty maxi-trials, all of which ended with convictions and heavy sentences (Paoli, 2003: 214).

Although they do not provide regional specification, the data of the Department of Penitentiary Administration confirm that the risk of spending a long time in jail has become real for *mafiosi*. On December 2005, for example, a total of 5,684 persons were serving sentences for the crime of mafia association foreseen by Article 416*bis* of the Italian Criminal Code, including 5,514 male Italians, 45 female

Italians, 177 male foreigners and 8 female foreigners (Ministero della Giustizia, 2006). Of those, 672 *mafiosi* were held under the special incarceration regime (Art. 41*bis* of the penitentiary law) as of the end of December 2002. The imprisoned chiefs were so discouraged that in early 2001 some of them allegedly proposed a deal to the state institutions: they would confess to their own crimes, without involving other mafia members, in exchange for a reduction in their convictions and the abolition of the special detention system.

The financial drain on the two organisations has been especially heavy. During the Olimpia-1 trial, for example, the Reggio Calabria court seized properties worth almost €80 million and confiscated goods worth over €20 billion definitively (*Gazzetta del Sud* January 20, 1999; see Paoli, 2003: 215). According to the prefect of Reggio Calabria, assets worth €700 million were seized in the province during the 1990s (CPM, 2000: 51). The DIA alone claims to have seized assets worth over €1 billion from Sicilian mafia groups. Whereas the precision of these estimates may be questioned,[3] qualitative evidence confirms the overall impact. Indeed, some mafia families now seem to be virtually bankrupt as a result of seizures and sentences. According to several law enforcement investigators, during the second half of the 1990s many mafia groups, both in Palermo and in Reggio Calabria, stopped paying the monthly salary to the families of the convicted "men of honour," thus ending one of the most important principles of the mafia legal order because they no longer have liquidity (DIA, 2007: 58).

The new investigations were frequently promoted by confessions from former mafia members, dozens of whom decided to become state witnesses after the early 1990s. At its peak, in late 1996, more than 1,200 former members of criminal groups were under the state protection programme. About 35% of them formed part of Sicilian crime coalitions. The percentage of *pentiti* coming from the ranks of Calabrian mafia and pseudo-mafia groups, which rely more than Cosa Nostra on family blood ties, was instead more modest (13%) (Ministero dell'Interno, 1997). The sudden rise in the number of mafia witnesses was made possible by the adoption of legislation granting them sanctioning and penitentiary benefits and establishing a state protection programme in 1991. However, the sudden rise in their number also reflects the crisis of the mafia legitimation system following the modernisation of southern Italy and the partial "entrepreneurial transformation" of mafia groups, which was aggravated by Cosa Nostra's more and more indiscriminate use of violence in the late 1980s and early 1990s (Paoli, 2003: 94–98).

Mafia witness confessions and statements allowed not only investigations and arrests but also inquiries into what is sometimes inappropriately called the "third level": namely the political and judicial protection mafia groups enjoyed for decades.

---

[3] The value of seized assets is estimated in a rough and approximate way by the local police sections at the moment of seizure. A careful evaluation of the asset value is carried out only when the confiscation decision is final and the assets have to be incorporated into the state property or sold by auction (see Paoli, 2002a).

The three most sensational investigations and trials begun in the early 1990s – namely, the two involving Giulio Andreotti,[4] accused in Palermo of belonging to a mafia-type delinquent association and in Perugia of having ordered the murder of journalist Mino Pecorelli in 1979, and the trial against Corrado Carnevale, the former head of the first section of the Corte di Cassazione – largely backfired, as the two defendants were finally acquitted of all charges. Andreotti's acquittal, however, was not stainless. The Palermitan judges in fact pointed out that Andreotti, together with some Sicilian Christian Democrats, had close relationships with Cosa Nostra before 1980, though they applied the statute of limitations for that period. Moreover, with less media outcry, several other politicians have been brought to trial accused of favouring mafia groups and some of them have been convicted — among them, Marcello Dell'Utri, Silvio Berlusconi's former right hand, who was convicted by a first-degree Palermitan court for his support for Cosa Nostra.

Cosa Nostra and the 'Ndrangheta's Reactions and Current Conditions

Cosa Nostra and the 'Ndrangheta have responded in several ways to escape law enforcement action or at least to minimise its impact. For Cosa Nostra this has meant a reversal of the open challenge to state sovereignty pursued in the early 1990s. The radical change of strategy has not been accompanied, however, by a significant change of leadership: Bernardo Provenzano who used to be Cosa Nostra's undisputed leader until his arrest in 2006, was Totò Riina's right hand in the 1980s and early 1990s.

The first goal of both organisations' reforming efforts has been to become as invisible and impermeable to law enforcement agencies as possible. In line with this objective, except for the anomalous murder of the vice-president of the Calabrian regional assembly, Francesco Fortugno in October 2004, no murders of civil servants or terrorist attacks have been authorised. Within Cosa Nostra, rigid procedures have also been established to authorise the murder by "men of honour" of other mafia members or petty criminals. This new policy has had tangible results, as Table 2.2 reveals. Since the early 1990s there has been a dramatic decrease in the number of murders, and specifically organised crime-related murders, recorded in Sicily and Calabria. In 2003 mafia murders committed in Sicily were one twenty-fifth of those recorded in the peak year of 1991, while in Calabria they fell to one sixth.

Allegedly, Provenzano has also asked his mafia brothers to focus on entrepreneurial activities that do not raise much social alarm, such as extortions, usury, manipulation of public tenders and, to the extent possible, drug trafficking. Changes are also recorded in the very organisation of extortions. According to the DIA, all producers,

---

[4] Giulio Andreotti is one of the most important politicians in the post-war period: he has been a member of parliament since 1948, prime minister seven times, and a government minister countless times.

Table 2.2 Murders and organised crime-related murders reported in Calabria and Sicily

	Calabria		Sicily	
	Murders	Org. crime murders	Murders	Org. crime murders
1990	326	141	428	150
1991	277	165	481	253
1992	151	46	399	200
1993	126	43	252	85
1994	121	42	249	90
1995	95	24	223	88
1996	103	30	180	66
1997	100	32	131	34
1998	85	28	140	35
1999	82	26	116	28
2000	84	34	86	13
2001	88	28	82	20
2002	61	17	70	11
2003	69	26	61	10

Source: ISTAT, Annuario Statistische Giudiziarie Penali, several years

and not just the big companies as in the past, are now asked to pay a contribution to Cosa Nostra, but contributions are kept low to prevent popular resentment from reaching critical dimensions (Ministero dell'Interno, 2006: 20–21).

To ensure cohesion and reduce the number of potential defectors, Provenzano, according to the DIA, also envisaged and implemented a fully-fledged plan with three main pillars:

- Returning to Cosa Nostra's traditional rules, which had in the past allowed the organisation to operate inconspicuously
- Drastically reducing the number of "men of honour," forming in fact a criminal elite set apart from "manpower," to create some layers of protection around "men of honour" and thus protect them from possible defectors among low-ranking staff
- Raising the cultural standard of Cosa Nostra leaders and members, by recruiting candidates with high educational standards and a good social position (Ministero dell'Interno, 2000)

To protect himself from possible defections, Provenzano also relied on a small number of trusted *mafiosi*, who were entrusted with responsibilities over and above the traditional organisations in families and districts (the so-called *mandamenti*) (Ministero dell'Interno, 2001b). According to several sources, even the Palermitan provincial commission (undoubtedly Cosa Nostra's most consolidated collegial body) has not held full meetings since 1994. Up until his arrest, Provenzano used to communicate primarily with his trusted lieutenants via so-called *pizzini*, i.e., short, hand-written or typed letters that were delivered back and forth by chains of trusted postmen. Ironically, this communication system, which had been dictated by security reasons, has now turned into an unexpected and very rich source of information and evidence against high-ranking Cosa Nostra members.

The *pizzini* themselves, as well as the numerous wiretappings of conversations among *mafiosi*, also attest to the economic and political decline of Cosa Nostra. While international drug trafficking continues to be an important activity for some 'Ndrangheta families, particularly those of the Ionian coast, Cosa Nostra groups are by now largely marginalised.

To illustrate how Cosa Nostra's role in drug trafficking has changed," the DIA wrote in one of its bi-annual reports, "it is enough to point out that the Bagheria family, which once had a prominent role in heroin trafficking towards the United States, has now turned to supplying a dealers' network in Palermo (Ministero dell'Interno, 2000: 11).

Extortions currently represent the primary and most stable source of revenue for most Cosa Nostra families. Thus, far from profiting from Cosa Nostra and the 'Ndrangheta's entrepreneurial difficulties, people and companies in high-density mafia districts are today exploited by the *mafiosi* more than ever, as they are called to make up for the failed earnings from drugs or other illegal trades.

All of the units of the two mafia confederations place their hopes for economic recovery in gaining public contracts, which began to be distributed from 2000, after the sharp drop, especially in the South, following the Tangentopoli (Bribesville) inquiries. Cosa Nostra and 'Ndrangheta families have been particularly eager to intercept part of the considerable sums from the EU funds of Agenda 2000: approximately €9,000 and €5,000 million, respectively, were invested between 2000 and 2006 in Sicily and Calabria. However, the extent to which the mafia groups have been able to achieve their objectives is unclear. A co-ordination committee for the supervision of major public works, composed of representatives of law enforcement agencies and several national Ministries, began to operate in 2003 and has so far failed to find evidence of large-scale mafia infiltration in major infrastructural investments. Indeed, even Provenzano's *pizzini* seem to prove Cosa Nostra's current inability to influence the allocation of large-scale projects in contrast to predominant and more successful involvement in more local and limited tenders (Loi, 2006).

Cosa Nostra and 'Ndrangheta families still enjoy a local network of trusted politicians, most of whom are members of the Unione dei Democratici Cristiani (UDC) or Forza Italia: it suffices to mention that Salvatore Cuffaro (UDC), the President of the Region of Sicily, is currently under trial for abetting Cosa Nostra. During the 5 years of Berlusconi's second and third cabinet (2001–2006), mafia bosses had also hoped that the many southern Italian lawyers who routinely represented *mafiosi* and had been elected to Parliament in the ranks of the government coalition would reform current antimafia policies. However, their hopes were largely frustrated, despite the not-so-veiled threats from imprisoned bosses. As Francesco Messina Denaro summed up in a *pizzino* to Provenzano, "we have been hit hard… if they [the law enforcement agencies] go on so, they will not only arrest all of us but even our chairs" (Cordella, 2006: 7). Messina Denaro went on asking Provenzano for political help, but he then added himself: "they [the politicians] do not do anything for nothing. And at the moment, we do not have much leverage over them" (Cordella, 2006: 7).

Concluding Remarks

The radical and painful changes undergone by the two principal sets of organised crime in southern Italy are the best proof of the effectiveness of antimafia law enforcement action. Cosa Nostra has seen all of its high-ranking members of the 1980s and 1990s arrested, has been excluded from most international trafficking and has lost political power nationally. The 'Ndrangheta is in a slightly better condition but many of its families, too, have been decimated by arrests, convictions and asset confiscations. However, the capacity of mafia groups to survive and regenerate should not be underestimated: Cosa Nostra, in particular, re-emerged strengthened from the two previous waves of sharp repression in the 1920s–early 1930s and during the 1960s. Recognising mafia groups' current crisis does not mean that they are defeated or that the law enforcement campaign against the mafia should be relaxed.

A brief look at the socio-economic context explains why a recovery cannot be ruled out. Mafia groups keep on finding a fertile breeding ground in the chronic underdevelopment and inefficient public government of some parts of southern Italy, which are usually worse where mafia and organised crime are the strongest. As far as underdevelopment is concerned, it suffices to say that for much of the past decade the unemployment rate in Calabria and some Sicilian provinces, such as Catania, Palermo and Enna, has oscillated between 28% and 35% and in 2005 was still as high as 18.6 for the whole Mezzogiorno (Svimez, 2006). Youth unemployment is even more dramatic. In Calabria, for example, youth unemployment (i.e., among 15–25 year olds) was as high as 66% in 1999, with a peak of 71% in the province of Reggio Calabria (CPM, 2000). Given these conditions, a career in the mafia or crime still looks attractive to many youngsters with poor education and few hopes of finding a job in the legal economy, and who thus provide an inexhaustible reserve army of criminal manpower. Much for Southern Italy is also poorly and ineffectively governed by plethora of local politicians and civil servants willing to spread resources to friends and patrons in order to buy consensus. Even when these state representatives are not directly linked to the mafia, they end up perpetuating old clientelistic logics in which mafia groups can hide and flourish. Despite the recent law enforcement successes in the fight against the mafia, therefore, the road ahead is still a long one and it would be a terrible mistake to claim premature victory.

References

Becchi, A., & Turvani, M. (1993). *Proibito? Il mercato mondiale della droga*. Rome: Donzelli.
Bolzoni, A., & Lodato. S. (1998). *C'era una volta la lotta alla mafia*. Milan: Garzanti.
Cordella, S. (2006). Matteo Messina Denaro: il re di Trapani. *Antimafia Duemila, 6*(3), 7–10.
CPM (Commissione Parlamentare d'inchiesta sul fenomeno della mafia e sulle altre associazioni similari). (2000). *Relazione sullo stato della lotta alla criminalità organizzata in Calabria*, doc. XXIII, no. 42, XIII Legislature. Rome: Camera dei Deputati.

Cressey, D. (1969). *Theft of the Nation*. New York: Harper and Row.

DIA (Direzione Investigativa Antimafia). (2007). *Rilevazioni statistiche*. Accessed January 2007. <http://www.interno.it/dip_ps/dia/pagine/rilevazioni_stat.htm>.

Franchetti, L. ([1876] 1993). *Condizioni politiche ed amministrative della Sicilia*. Reprint, Rome: Donzelli.

La Spina, A. (2004). The paradox of effectiveness: Growth, institutionalisation and evaluation of anti-mafia policies in Italy. In C. Fijnaut & L. Paoli (Eds.), *Organised crime in Europe: Concepts, patterns and policies in the European Union and beyond* (pp. 641–676). Dordrecht: Springer.

Lestingi, F. (1884). L'associazione della Fratellanza nella provincia di Girgenti. *Archivio di psichiatria, antropologia criminale e scienze penali, 5*, 452–463.

Lodato, S. (1999). *Ho ucciso Giovanni Falcone. La confessione di Giovanni Brusca*. Milan: Mondadori.

Lodato, S. & Grasso, P. (2001). *La mafia invisibile. La nuova strategia di Cosa Nostra*. Milan: Mondatori.

Loi, M. (2006). In carcere il Gotha di Cosa Nostra: I pizzini di Provenzano svelano i nuovi capi. *Antimafia Duemila, 6*(3), 2–6.

Ministero dell'Interno. (1997). *Relazione sui programmi di protezione, sulla loro efficacia e sulle modalità generali di applicazione per coloro che collaborano alla giustizia – II semestre 1996*.

Ministero dell'Interno. (2000). *Relazione semestrale sull'attività svolta e i risultati conseguiti dalla Direzione Investigativa Antimafia nel secondo semestre del 1999*.

Ministero dell'Interno. (2001a). *Rapporto sullo stato della sicurezza in Italia 2001*.

Ministero dell'Interno. (2001b). *Relazione semestrale sull'attività svolta e i risultati conseguiti dalla Direzione Investigativa Antimafia nel secondo semestre del 2000*.

Ministero dell'Interno. (2005). *Rapporto sullo stato della sicurezza in Italia 2005*.

Ministero dell'Interno. (2006). *Relazione semestrale sull'attività svolta e i risultati conseguiti dalla Direzione Investigativa Antimafia nel secondo semestre del 2005*.

Ministero della Giustizia, Dipartimento dell'Amministrazione Penitenziaria. (2006). *Rapporto sulla situazione degli istituti penitenziari per l'anno 2005*.

Paoli, L. (2002a). Italien. In M. Kilchling (Ed.), *Die Praxis der Gewinnabschöpfung in Europa* (pp. 239–295). Freiburg: Edition iuscrim.

Paoli, L. (2002b). The paradoxes of organised crime. *Crime, law and social change, 37*(1), 51–97.

Paoli, L. (2003). *Mafia brotherhoods: Organized crime, Italian style*. New York: Oxford University Press.

Paoli, L. (2004). Organised crime in Italy: Mafia and illegal markets – Exception and normality. In C. Fijnaut & L. Paoli (Eds.), *Organised crime in Europe: Concepts, patterns and policies in the European Union and beyond* (pp. 263–302). Dordrecht: Springer.

Pezzino, P. (1987). Stato violenza società. Nascita e sviluppo del paradigma mafioso. In M. Aymard & G. Giarrizzo (Eds.), *La Sicilia* (pp. 905–984). Turin: Einaudi.

Scarpinato, R. (1992). 'Mafia e politica' in *Mafia. Anatomia di un regime*. Rome, Librerie Associate.

Svimez. (2006). *Rapporto 2006 sull'economia del Mezzogiorno – Sintesi*, 11 July 2006.

Weber, M. ([1922] 1978). In G. Roth & C. Wittich (Eds.), *Economy and society*. Berkeley: University of California Press.

Chapter 3
Hashish Smuggling by Bedouin in South Sinai

Emanuel Marx

Introduction

This paper examines the changing fortunes of the smuggling operations of South
Sinai Bedouin. The Bedouin are a link in the international drug traffic delivering
hashish and other drugs to the inhabitants of the Nile Valley. The full-scale entry of
the South Sinai Bedouin into drug smuggling began around 1950, and in less than
two decades smuggling grew into a major industry. At its zenith it provided about
30% of the aggregate income of the Bedouin population. Then smuggling stopped
almost overnight, for during the 15 years of Israeli occupation, from 1967 to 1982,
the crossing from the Sinai Peninsula into mainland Egypt was too dangerous for
the operators. During my fieldwork in South Sinai, which overlapped with the
Israeli occupation, most of the Bedouin were working as migrant laborers and a
handful entered the budding local tourist industry, so that the loss of income from
smuggling did not cause them economic hardship. But the leaders of the smuggling
gangs remained at home, and it was quite easy to meet them. They appeared to be
inactive, but I soon realized that they were working hard at keeping the smuggling
organizations alive: they fostered the ties with members of their former gangs and
looked after their mountain retreats. They were convinced that the political and
economic situation would sooner or later change, and that drug smuggling would
once again become feasible. Other Bedouin too acted to forestall an uncertain
future: they maintained orchards and small flocks, which at that time yielded no
income, as an economic reserve. They would fall back on it when migrant labor
would no longer be plentiful.

Conditions in Sinai changed again when it reverted to Egyptian rule in 1982. The
State developed hotel tourism as the major industry. The hotels employed only a
small number of Bedouin men, and the formerly thriving Bedouin guest lodges lost
some of their customers. Therefore many Bedouin returned to migrant labor, and
also increased their flocks and cultivated their orchards more intensively. The hashish
trade too revived very quickly, this time with a difference: now the Bedouin smugglers not only conveyed drugs to mainland Egypt, but also sold them to international
tourists and to fellow Bedouin. The drug dealers and other men were gradually
becoming smokers.

D. Siegel and H. Nelen (eds.), *Organized Crime: Culture, Markets and Policies.*
© Springer 2008

Why Drug Smuggling?

This story raises an intriguing question: how did hashish smuggling so rapidly become a major and seemingly permanent feature of the South Sinai Bedouin economy? The answer, I believe, is connected to three aspects of life in Sinai:

First, the Bedouin consider the smuggling of hashish and other drugs to be a legitimate economic enterprise, and not a criminal activity. Every Bedouin man spends part of his life in Egyptian cities, where the smoking of hashish is considered a harmless popular indulgence. Hashish had been smoked for centuries and the attitude toward it did not change even when the European imperialists had it outlawed. Furthermore, the drug is purveyed by powerful international organizations, and the Bedouin are proud to play a significant part in a great commercial enterprise. While the international drug trade mostly benefits big operators, in the Sinai leg of the operation, at least, the earnings were distributed widely among the many people who worked for it in various capacities, such as organizers, cameleers, truck drivers, boatmen, messengers, informers and collaborators from the ranks of government (cf. Levi, 1987: 373–374). The Bedouin respected and honored the chiefs of smuggling gangs, as entrepreneurs who had brought them a higher standard of living. Their attitude toward the drug trade was even tinged with defiance against the State which treated them as second-class citizens, whose agents suppressed and humiliated them, took over their land and deprived them of their livelihood, and whose cooperation often had to be bought (cf. Kershaw, 2006 for a similar situation in American Indian reservations).

Even the illegality of the hashish trade did not deter the Bedouin. They knew that they were risking arrest and imprisonment, but did not see this in a different light from ordinary businessmen who routinely take calculated risks. As modern people with a cosmopolitan outlook, they treated smuggling as a strictly rational business matter and, just like other businessmen, they "silenced morality" (Bauman, 2000: 29), as not being part of rational conduct. The prospect of higher earnings persuaded the smuggling chiefs in the early days of smuggling to include in some consignments heavy drugs, such as opium and heroin. From the 1980s onwards they became less selective about the goods they carried, as long as they made a good profit.

Second, the Bedouin treated the drug trade as a windfall that had to be exploited while it lasted. They could do so because they had built up a system of mutual insurance that allowed them to overcome setbacks and calamities. Repeated experiences of political and economic upheavals had taught them that misfortunes such as death, illness, imprisonment and impoverishment were part of life, were bound to happen and should be prepared for. In order to cope with these certainties, they provided the members of their households with a comprehensive system of social assurance, and devoted much time and effort to its implementation. Thus they maintained orchards and flocks that could become alternative sources of income if the need should arise, fostered networks of relationships by frequent visits to all members, and kept up a good relationship with God by regular personal and tribal pilgrimages to saints' tombs. The system gave the Bedouin not only the resilience to adapt to the rapidly changing circumstances of the drug trade, but also permitted them to accept with equanimity the cessation of smuggling in 1967. Today it allows them to face the daunting thought that smuggling may eventually end.

Third, the requirements of the smuggling trade fitted easily into the Bedouin's existing social realities and practices, such as their familiarity with cities, the wide dispersal and unceasing mobility of the Bedouin labor migrants, the free access of all tribesmen to pastures anywhere in South Sinai, the Bedouin's reliance on dispersed networks of kinsmen and friends and the absence of corporate agnatic groups, and the existence of lodgings (*maq'ad*) attached to the 160 odd saints' tombs spread along the east–west passages through the peninsula. Before developing the argument, I should explain the historical background of the drug trade.

Drug Smuggling in Historical Context

Since the thirteenth century the smoking of hashish has been one of the favorite pleasures of the Islamic world (Levey, 1971; Rosenthal, 1971). The practice was especially prevalent in Egypt, which used large quantities of locally produced hashish. Lane reports in 1836 that "The habit is now very common among the lower orders in the metropolis and other towns of Egypt" (Lane, 1895: 347). Some 150 years later, in 1990, hashish is still "the most commonly used drug in Egypt" (Hussein, 1990: v). But there is a fundamental difference between the two periods: while Indian hemp (*Cannabis sativa*), the raw material of which hashish is made, had for many years been grown locally and sold freely and cheaply, it became illegal as European involvement in Egypt increased. Under pressure from the European powers Egypt outlawed hashish production in 1879, with the result that it was from then on imported from Morocco and Greece.

Over the years the measures taken against drug use became ever harsher, to the point when Law 182 of 1962 made the "cultivation, smuggling, sale and distribution of restricted drugs felonies punishable by life imprisonment" (Hussein, 1990: v). The gradual criminalization of the drug trade could well be part of an encroaching Western imperialism. Agents of the state would justify the banning of the local drug hashish by its debilitating effect on the work and sexual prowess of users. At the same time they encouraged the popularization of the Western drug tobacco which then was believed to be quite harmless (Fournier, 2002: 12 shows that the French colonial authorities in Morocco adopted this policy at about the same time). It is no coincidence that tobacco and salt became government monopolies (Dumreicher, 1931: ix); both commodities were considered basic necessities of life and thus could become a perpetual source of government revenue. Salt was quarried by Bedouin from local deposits and the Egyptian Treasury used every means to wrest the trade from their hands. Tobacco and tobacco products were imported from Europe through Customs in the port of Alexandria. The third commodity, the harmful drug hashish, was to be banned and eliminated.

The Egyptian Coastguard Administration was set up in the 1880s to prevent the free trade in the three commodities. By the end of the century the Administration had branched out into three departments. One of these was the Desert Directorate, which became responsible not only for the prevention of drug smuggling but also for public security in the Libyan, Eastern and Southern deserts and the Sinai Peninsula (Dumreicher, 1931: xi). As the authorities knew that Egyptians did not consider hashish a harmful substance, they based the Desert Directorate chiefly on foreign personnel, European officers and Sudanese soldiers and non-commissioned officers, while "Egyptian sergeants ... were responsible for the clerical work" (Dumreicher, 1931: 6–7).

While the Egyptian Desert Directorate battled with the drug traffic, it also had a stake in maintaining the flow of drugs, so that it could continually make successful seizures. Individual law officers were always ready to enter into arrangements with smugglers which could augment their low salaries. The smugglers, on their part, often assisted the law enforcement agencies in discovering small shipments (cf. Dumreicher, 1931: 191), as it helped them to keep up the price of drugs. They also had a material interest in collaborating with the authorities, as this could assure the safe passage for their goods. These parallel interests could easily develop into permanent or intermittent collaboration between law enforcement agents and smugglers. Several government officials recorded in their memoirs instances of such collusion (Dumreicher, 1931: 191; Jarvis, 1931: 199; Russell, 1949: 275).

The outlawing of the drug did not affect consumer demand, but rapidly reduced local production. Therefore the price of the drug soon rose to the point where it became worthwhile for international businessmen to enter the lucrative trade. Most of the imports came overland or by sea from Morocco. Then Greek businessmen entered the fray. Their hashish was grown and produced in Greece, transported by boat to desolate spots on the Egyptian Mediterranean coast, where camel caravans took over and carried the merchandise into the Nile Valley. Only when Greece prohibited the cultivation of hashish in 1932, Syria and Lebanon rapidly took over (Russell, 1949: 271), without ever displacing Morocco as the biggest supplier of hashish to Egypt and to the world (United Nations Office on Drugs and Crime, 2005: 83).

Needless to say, tobacco has not displaced hashish, just as it has not reduced the use of qat and other local drugs in the Middle East. The imported and the local drugs are used side by side, sometimes simultaneously. Thus each participant in a Yemeni qat session provides his own bunch of qat to suit his means, while the host's tobacco pipe is passed round the circle of guests in an egalitarian manner (cf. Weir, 1985: 119). It is an irony of history that today the tables are turned: Western Europe has become the world's largest market for hashish and Middle Eastern countries are again major producers and exporters. According to a recent United Nations report "Western Europe is the largest market for cannabis resin, responsible for nearly 70% of global seizures in 2003, and 80% of this hashish was produced in Morocco" (UNODC, 2005: 81).

Smuggling of narcotics had interested the Sinai Bedouin since the beginning of the century (Dumreicher, 1931: 204; Russell, 1949: 272), but really took off in the 1930s, when the production of hashish in Syria and Lebanon accelerated. At first, the traders sent the drugs through Palestine and North Sinai into the Nile Valley. When the Customs control along the Palestinian border became more effective and the law agents apprehended many shipments, they began to shift to the longer but safer route through Jordan, Saudi Arabia, and across the Gulf of Aqaba and South Sinai into the Nile Valley. Only then did the Bedouin of South Sinai get involved in the drug trade. After the establishment of Israel in 1948 the North Sinai border became too dangerous, and the route through South Sinai became for nearly 20 years, until the Israeli occupation of Sinai in 1967, a major import channel of hashish into Egypt. At its zenith, the drug trade provided about 30% of the income of the Bedouin population. In retrospect, it is hard to comprehend how in just two decades it became such a major component of the economy of South Sinai.

During the Israeli occupation smuggling was at low ebb, because the Israeli and Egyptian forces were facing each other along the land and sea frontiers. Smuggling became too dangerous in these conditions. At the same time there was abundant work for labor migrants, so that most of the male adults were away at work. Some of the men who stayed at home were former smugglers who were now living in what Lavie (1990: 157) described as semi-retirement. Among them were some of the most prominent and interesting people I met during my fieldwork in Sinai. As their businesses were inactive at the time, they reluctantly agreed to talk about them. These men entertained a stream of visitors, maintained links with numerous kin, established contacts with the military and civilian authorities, and saw to it that the oases that served them as hideouts were kept alive. When the Israeli authorities permitted the resumption of the Mecca pilgrimage in 1976, the smuggling chiefs were among the first pilgrims, presumably in order to refresh their links with the leaders of the hashish trade.

The smugglers were not simply waiting for Israel to evacuate Sinai; they hoped to find ways and means to resume their activities under the watchful eyes of the Administration. Here is one example. One of the chiefs of the Muzaina tribe, formerly a prominent organizer of smuggling gangs, received a visit from the Israeli Military Governor of South Sinai. During their conversation the chief made a most humane offer: to convert a light truck into a hearse and to bring back for burial the bodies of Bedouin who had died in Israeli hospitals. He volunteered to pay for all the expenses. The Israeli Administration used to fly seriously ill and wounded Bedouin for treatment at the Tel Hashomer hospital near Tel Aviv. The Administration had neither buried the bodies of those Bedouin who had died in hospital nor found a way to return them for burial in Sinai. Intuitively I understood that the chief had devised an ingenious way of importing drugs from Israel, as no customs officer would search a decomposing body. I assume the Military Governor too knew that he was being asked to connive at a renewal of drug smuggling. But with diplomatic politeness he assured the chief that he would consider the request favorably.

The Organization of Drug Smuggling

The smuggling gangs recruited some members on a kinship basis and others according to the services they could provide. The nuclei of such enterprises were often sets of brothers, whose abodes were spaced out along a west-east axis, presumably to facilitate the movement of goods. As each member also brought in maternal and affinal relatives, as well as unrelated persons who fulfilled useful functions, as boatmen or truck-drivers, or because they had useful ties with government agents, eventually many members of gangs were not related by kinship. This was not unlike the chain migration of laborers in which each man, in his turn, joins a relative or friend who is already established in a particular place of work. Such an arrangement assured both the loyalty of the members and the relatively wide and even distribution of the gang's earnings. As the members of smuggling gangs were recruited from diverse tribes, the smuggling operations were not hampered by territorial claims. They were analogous to the herding arrangements: just as flocks were permitted to graze anywhere in the South Sinai, so smuggling gangs moved all over the region. But they kept away from villages and tent camps where they ran the risk of exposure. Instead, they tended to travel by night and to take breaks at saints' tombs. The 160 or so saints' tombs [see a list and map in Levi (1980: 139–148)] are conveniently strung out along the main routes that cross the mountainous interior from east to west. The tombs are located close to water sources, and there are huts for pilgrims who wish to stay overnight. These are equipped with cooking utensils and a small stock of food that each visitor is expected to replenish. As all the tombs are frequented by bona fide pilgrims throughout the year, they provide a suitable alibi to smugglers.

The earnings from drug smuggling filtered down from the organizers to the rank and file of their employees. Smuggling thus made an important contribution to the economy of South Sinai. It was the driving force behind the development of orchards, the digging of new wells and installation of motorized water pumps, and the development of new oases which served as hideouts for smuggling gangs.

These oases were located in relatively inaccessible locations in the mountains. They were created from scratch, even where ecological conditions were not ideal. Due to the irregularity of drug transports, there would often be months of inactivity between hauls. The smugglers would spend them in their mountain retreats, away from prying eyes. The leaders of smuggling rings and their henchmen planted small but intensively cultivated orchards which, in addition to the usual almonds, pomegranates, apples and quinces, often included varieties not commonly found in South Sinai, such as mandarine, tangerine, plum and avocado. They were keenly interested in introducing new varieties, as they aimed to spread the supply of fresh fruit over the longest possible period. In the 1950s, when smuggling became a major source of income for the South Sinai Bedouin, the entrepreneurs made further investments in these oases. They spent a great deal of money on diesel engines and pumps, in order to make irrigation less arduous. They also built solid homes for their families, on the expectation that they would often be absent on drug hauls. When the drug traffic ceased during the Israeli occupation, the oases rapidly fell into desuetude.

Bir 'Oqda and Bir Zrer are good examples of this type of oasis. In the early 1970s, the population in Bir 'Oqda rapidly declined, and most of the twenty odd households moved to Dahab on the coast. In former years they had gone to Dahab in the late summer, to enjoy the breeze and supplement their diet with sea food, and wait for the date harvest at the end of August. Around October they used to return to the mountain village. Now they simply stayed on in Dahab and delayed their departure from day to day. Only a few men made short trips to Bir 'Oqda, where the water pumps fell into disrepair and the fruit trees dried up. They explained the situation in ecological terms: the water sources in Bir 'Oqda were not as copious as they used to be, and therefore the yield of the fruit trees was dwindling. There simply was no point in cultivating the oasis. But another explanation may be more relevant: during the occupation, the smuggling of narcotics to Egypt stopped almost completely. Therefore many former smugglers felt that the hideout was no longer required; they saw no reason why they should invest more resources in the construction of new wells and in improving the orchards. They became labor migrants like the rest of the male population. Still, from time to time they returned to Bir 'Oqda to look after their orchards. They often argued that they wished to postpone the inevitable decline of the oasis, and as often thought that they had to maintain it in a state of readiness for a possible revival. The big operators looked at the situation quite differently. They hoped and expected conditions to change and had the resources to hold out till then. Therefore they spent much of their time in the oasis and hung on to their orchards. In the meantime, they lived on their often quite substantial hoarded savings and regularly entertained guests, among them many associates from their smuggling days.

The inhabitants of Bir Zrer responded differently to the changed situation. A new highway from Eilat to Sharm al-Sheikh, constructed by the Israeli administration, passed close to their village. They built a track to the main road, invested money in trucks and pickups, and became involved in transporting migrant workers and tourists. The population of the village actually grew. This does not mean that they discarded the smuggling option. If the tourist trade declines or if they are pushed out by the big hotels, they will once more resort to labor migration and smuggling.

The new-found wealth from smuggling also transformed the larger palm oases, especially the central oasis in Wadi Firan. It has always been the major east-west thoroughfare, and almost all the traffic, including smuggling caravans, passes through it. Members of the various tribes owned shares in its palm trees. Now they also built solid houses and set up vegetable plots in the oasis, so that it became exterritorial. Here, too, mechanical pumps were installed. One outcome of the new situation was that the water level in the oasis gradually fell. In consequence the wells had to be deepened and became more expensive to maintain. The shallower wells downstream dried up and only wealthy Bedouin could afford to carry out the required excavations. The poorer Bedouin moved out of the palm groves into the vicinity of perennial wells upstream. The trend of oasis society to become ever more exclusive ended, however, under the Israeli occupation, when wage labor became plentiful and almost all the men took part in it and earned good wages. Now every owner of an orchard could afford a motor pump and install plastic water pipes, and the downward spiral of declining ground water levels and ever deepening wells accelerated.

Another outcome of drug smuggling was the self-declared peacefulness of the South Sinai Bedouin. Men often pronounced: "We Bedouin are a peace-loving people who never engage in fighting. In 50 years there has not been a single case of murder". They also pointed out that, in contrast to other regions, men did not carry weapons. These categorical statements referred to the need to maintain the public peace throughout the Peninsula, as a precondition for large scale smuggling. On their way over the mountains, the drugs crossed the territory of several tribes. The safe passage of drugs could be assured only if they all agreed to collaborate for a common good and drop their differences. This became possible because so many members of the various tribes participated in the drug trade and the whole population benefited from the wealth it created.

In a sense, the peaceful atmosphere continued during the period of my field-work, a time when smuggling had ceased. South Sinai was very tranquil and safe and Bedouin men and women, as well as tourists, moved freely and without hindrance over the countryside. The few violent incidents that did occur, involved former smugglers. For instance, a smuggler who had spent more than a decade in Egyptian prisons was shot at by an old competitor and lost the fingers of one hand. Now that smuggling was in abeyance the men could settle outstanding accounts. However, it appears that when the hashish trade resumed in 1982, all these disputes were put aside and peace reigned once more.

Future Prospects

Sinai has become an important tourist center. Although the hotels have been built on land taken from the Bedouin, they employ only a small number of Bedouin in menial jobs (Aziz, 2000: 32). Thus migrant labor has again become the main source of income. The transportation of drugs through South Sinai into mainland Egypt has also regained its place in the Bedouin economy. In addition, some Bedouin make a living by selling drugs to tourists and have themselves become consumers. As to the future, one may safely predict that hashish smugglers will sooner or later deal in even more profitable outlawed and heavily-taxed commodities, such as weapons, diamonds, and addictive drugs like heroin and tobacco. Then these operators will become so wealthy and powerful that they will dispense with the moral approval of their fellow citizens. At first, they will carry them along on the expectation of higher rewards, and later – coerce them into submission. Ultimately, this may lead to a situation in which the smugglers operate quite openly, work hand in glove with some local and national authorities, employ the most technically advanced means of transportation and communication, and deal in a wide and variable range of contraband commodities, just as they do in the Sahara today (Claudot-Hawad, 2006: 673–674). But not only government agencies may be involved in the trade. Militant political groups may gain control over existing smuggling organizations or set up their own, and use them to finance their activities, which may include acts of terror. That seems recently to have happened in Sinai, where in 2005 for the first time Bedouin took part in terrorist attacks on hotels in Sharm al-Sheikh City and Taba. The discontent of two groups, the

Bedouin and the militant political groups (often ignorantly called "Muslim funda-mentalists"), converges in these terrorist acts. The Bedouin protest against the State that has handed over their land to the hotel chains, and against the hotels that monopolize tourism and employ only few Bedouin. And the militants attack the State that does not allow them a share in political power. They attack the State at the point where it hurts most: for each assault on foreign tourists interrupts the flow of tourists for many months, causing irreparable losses of revenue.

There is no easy way to end the traffic in drugs. As long as South Sinai remains the safest route into Egypt, it will retain its attraction to the international business world, and the traffic of contraband goods will continue. Even if the State were to legalize the production and consumption of hashish, and provide the Bedouin with regular employment, this state of affairs would persist (cf. Bovenkerk, 2004). Smuggling would continue unabated, punctuated by recurrent waves of terrorism. The State could, of course, declare war on the drug traffic in Sinai, and cause it to be diverted to safer routes. Such a course of action would deeply affect the economy of the Sinai Bedouin, but would not stem the flow of hashish and other contraband to Egypt.

References

Aziz, H. (2000). Employment in a Bedouin community: The case of the town of Dahab in South Sinai. *Nomadic Peoples, 4*(2), 28–47.

Bauman, Z. (2000). *Modernity and the Holocaust.* Ithaca: Cornell University Press.

Bovenkerk, F. (2004). The dark side of Dutch drug policy or the failure of half-way legalization. In *Global drug policy: Building a new framework* (pp. 161–164). Paris: Senlis Council.

Claudot-Hawad, H. (2006). A nomadic fight against immobility: the Tuareg in the modern state. In D. Chatty (Ed.), *Nomadic societies in the Middle East and North Africa: Entering the 21st century* (pp. 654–681). Leiden: Brill.

Dumreicher, A. von. (1931). *Trackers and smugglers in the deserts of Egypt.* London: Methuen.

Fournier, G. (2002). *Global drug policy: A historical perspective.* Paris: Senlis Council.

Hussein, N. H. (1990). *The sub-culture of Hashish users in Egypt: A descriptive analytic study.* Cairo Papers in Social Science 13[2]. Cairo: American University in Cairo Press.

Jarvis, C. S. (1931). *Yesterday and to-day in Sinai.* Edinburgh: Blackwood.

Kershaw, S. (2006). Tribal underworld. *New York Times,* February 19–20, 2006.

Lane, E. W. (1895). *An account of the manners and customs of the modern Egyptians.* London: Gardner. (First published in 1836).

Lavie, S. (1990). *The poetics of military occupation: Mzeina allegories of Bedouin identity under Israeli and Egyptian rule.* Berkeley: University of California Press.

Levey, M. (1971). Hashish. *Encyclopaedia of Islam, New Edition, 3,* 286–287.

Levi, S. (1980). *Emuna vepulhan shel habeduim biderom Sinai* (Belief and ritual among the bedouin in South Sinai). Tel Aviv: Society for Protection of Nature, Tsukei David Field School.

Levi, S. (1987). *Habeduim bemidbar Sinai* (The bedouin in the Sinai Desert). Tel Aviv: Schocken.

Rosenthal, F. (1971). *The herb: Hashish versus Medieval Muslim Society.* Leiden: Brill.

Russell, T. (1949). *Egyptian service: 1902–1946.* London: Murray.

United Nations Office on Drugs and Crime. (2005). *World drug report 2005.* Accessed June 2007. <www.unodc.org/unodc/en/world_drug_report.html>.

Weir, S. (1985). *Qat in Yemen: Consumption and social change.* London: British Museum Publications.

Transnational Flows

Chapter 4
Corruption as a Causal Factor in Human Trafficking

Sheldon X. Zhang and Samuel L. Pineda

Introduction

Human trafficking has gained wide attention in both developed and developing countries. It is now considered the third-largest criminal industry in the world as of 2006 after the arms and drugs trades, with profits reaching billions each year (Siobhan, 2006). According to FBI estimates, human trafficking generates $9.5 billion annually in the U.S. alone (Siobhan, 2006). It is a phenomenon that shows no prejudice toward race, gender, or geography, but a general preference towards the young and female. Trafficking victims are a diverse group, ranging from those who are aspiring to improve their life to those desperate to escape civil unrest. Some leave their home countries on their own, while others, often children and women, are "given" or sold to related or unrelated adults who promise education and employment opportunities. Invariably, they fall victim to forced or bonded labor.

Human trafficking has found supporters throughout the world who hold fundamentally differing views towards the rights and welfare of a human being. Estimates of the number of people trafficked each year vary tremendously, and there are few empirical sources for making such figures valid and reliable. Kevin Bales, a sociology professor from Roehampton University in London and one of the few researchers who conducted field research and sought to gauge the extent of modern day slave labor worldwide, estimated that approximately 27 million slaves exist globally (Bales, 2005). The U.S. State Department's annual Trafficking in Persons Report (or the TIP Report) estimates that some 12.3 million people are enslaved in forced labor, bonded labor, sexual servitude, and involuntary servitude at any given time around the world. Although leading the battle against transnational human trafficking, the U.S. also has its own share of the problem (U.S. Department of State, 2005). For instance, on December 17, 2005, Attorney General Alberto Gonzales announced the breakup of a large prostitution ring involving more than 30 children as young as 12 having sex at truck stops, hotels, and brothels (Associated Press, 2005). Multiple indictments were made against traffickers in states including Michigan, New Jersey, and Pennsylvania for bringing children across the state lines for prostitution. Trafficking victims typically are transported into the U.S. from Mexico to the southern states, and from Asia to California, and by the Russian mafia into New York area.

Challenges of Studying Human Trafficking

The problem of human trafficking touches just about every country in some aspect. There are sending countries where prospective victims are recruited and transported, transit countries that serve as way stations, and destination countries where vice lords eagerly await fresh supplies of women and children. The flow appears to be from less developed nations, such as Mexico and Latvia, to more affluent ones such as the United States and Germany. This observation has led researchers to believe that human trafficking is causally related to poverty. This theory has been echoed by many governments and non-government organizations. Based on information obtained from police reports and interviews with victims in shelters, researchers in many parts of the world have come to accept the explanation that these women and children are primarily driven by poverty into slavery. The problem with the poverty assumption is that it has been accepted mostly on face value with limited empirical scrutiny and to the exclusion of other factors.

The successful implementation of anti-trafficking policies begins with a clear definition of the problem. For many years, the term *human trafficking* was used interchangeably with *human smuggling* with little differentiation between the two concepts. It is now accepted now that the phrase *human trafficking* (or trafficking in person) specifically refers to not only the surreptitious entry of people but also the use of force, fraud, coercion, and violence during the process of transportation for the purpose of subjection to involuntary servitude, peonage, debt bondage, or slavery. This definition was established in December of 2000 with the signing in Palermo, Italy of the *UN Protocol to prevent, suppress and punish trafficking in persons, especially women and children, supplementing the United Nations convention against transnational organized crime.*

Prevent, Suppress, and Punish Trafficking in Persons

"Trafficking in persons' shall mean the recruitment, transportation, transfer, harboring or receipt of persons, by means of threat or use of force or other forms of coercion, of abduction, of fraud, of deception, of the abuse of power or of a position of vulnerability or of the giving or receiving of payments or benefits to achieve the consent of a person having control over another person, for the purpose of exploitation. Exploitation shall include, at a minimum, the exploitation of the prostitution of others or other forms of sexual exploitation, forced labor or services, slavery or practices similar to slavery, servitude or the removal of organs" (United Nations, 2000: 2).

This current definition of human trafficking has evolved over a long period of time and with much debate. By most accounts, victims in trafficking are primarily for labor exploitation. A study by the International Labor Organization (ILO) estimated that less than 10% of the millions of victims of forced labor in Asia are trafficked for commercial sexual exploitation (Feingold, 2005: 26). As a result, the ILO has constructed its own definition of human trafficking to emphasize the slavery of labor persuasion: (1) the work or service is extracted under the menace of penalty and it is (2) undertaken involuntarily" (International Labor Organization, 2005). Because of its shock value, trafficking for commercial sex often commands a larger audience in the political discourse. Feminists in early twentieth century were active in lobbying government agencies in the western countries to abolish brothels and end all forms of prostitution (Outshoorn, 2005). Their successful political activism eventually tied prostitution to trafficking and subsequently led to the passage of United Nations' *International Convention for the Suppression of the Trafficking In Persons in 1949* that called on all states to "suppress not only trafficking but also prostitution, regardless of whether or not they occur with the consent of the women involved" (Outshoorn, 2005: 142). This view was later modified with the arrival of another group of feminists who "sees prostitution as a possible option or a strategy of survival taken by women, which should be respected" (Outshoorn, 2005: 145). Despite its wide acceptance, the interpretation and policy operation of the UN definition on human trafficking remains a messy process and the term continues to be used to describe various groups of people (Gozdziak & Collett, 2005: 106).

   In addition to definitional challenges, the extent of human trafficking remains largely a guess work. Empirical data are hard to obtain for obvious reasons. But despite the secretive nature of the business, many government and non-government agencies somehow have managed to come up with estimates, most of which are nothing more than wild estimates. The U.S. government, for instance, has for years produced various numbers for estimating the extent of human trafficking within the U.S. In 1999, the U.S. government officials offered a count that ranged from 45,000 to 50,000 victims trafficked into the U.S per year. Those estimates changed to a range of 18,000–20,000 in 2003 and dropped further to the current range of 14,500–17,500, as stated in the 2005 TIP report (U.S. Department of State, 2005). Little information was provided to disclose the methodological procedures for arriving at these estimates, let alone an explanation of the precipitous drop over such a short period of time in the absence of any successful social or law enforcement campaigns.[1]

   Kevin Bales (2005) compiled a list of estimates for countries with slave labor. His numbers differ in many cases significantly from those of other sources. For example, the State Department (2005), along with the Central Intelligence Agency, has estimated that the number of people trafficked to the United States ranges from 14,500 to 17,500 each year. Bales estimates that "currently there are between 100,000 and 150,000 slaves inside the U.S." (Bales, 2005: 185). According to his estimates, India and Pakistan led the world in the number of trafficking victims.

---

[1] Some discussion was provided in recent years on the use of Bayesian models to estimate the number of trafficking victims in the U.S. Statistical procedures such as the Bayesian modeling can be used to produce estimates when missing values are present in the primary data. However, the estimates are suspicious when none of the missing values or the sources of the primary data were discussed in the report. To date, no explanation has been offered on how these estimates have been compiled over the years and reasons for the sharp downward trend.

India was estimated to have 20 million slaves; Pakistan 3 million. The next closest countries for comparison were Mauritania, Nepal, and China, each having an average of 275,000 slaves per country. The majority of the remaining countries had a range of 10,000–20,000 slaves.

In recent report, the ILO estimated that globally there are at least 12.3 million people in forced labor, 11% of which involve sexual exploitation (Belser, Cock, & Mehranl, 2005). Unlike most other estimates, the ILO report provided a detailed account of the methodology used in the estimate and evaluated the results against those of other sources. For instance, the ILO methodology included two main steps: an estimate on the global number of reported cases of forced labor and the total number of reported victims, based on a so-called "capture-recapture" sampling method and leads to an estimate of total reported victims; and secondly, an estimate of the actual number of persons in forced labor in major forms of forced labor such as commercial sex and economic exploitation (Belser et al., 2005: 14).

Changes in the numbers of trafficked victims can result from different interpretations of the definition that researchers apply when gathering data. Human trafficking ranges from sexual exploitation to non-sexual coerced labor, for instance, labor as a form of payment for a collateral debt. Some of the victims in such indentured labor last a short period of time while others appear to be in a lifelong bondage. Some examples described by Masci (2004) include a village boy seized by Arab raiders and forced to herd livestock until he escaped some 10 years later, young women from Macedonia responded to an advertisement for work in Italy but found themselves forced into prostitution, and children being sold for percentages in returns by their parents into pedophile rings in Thailand. To complicate the problem further, estimates on the extent of human trafficking vary because of the political persuasion of the organization compiling the data. For instance, NGOs tend to inflate numbers in an attempt to force governments to take action (Bales, 2005).

Explaining Human Trafficking

A common explanation of human trafficking is poverty. Many researchers have grown accustomed to using poverty and its concomitant factors such as lack of education and job opportunities to explain why women and children fall prey to human traffickers, particularly for those of eastern European origins. In his analysis of the aftermath of the breakup of Balkan war, Rathgeber (2002) found that women in Romania, Moldova, and other countries were driven by poor social and economic conditions and allured by promising advertisements for overseas opportunities. Traffickers take advantage of these women's desperation and lure them into or through countries that do not have the support services or law enforcement capabilities to prevent such abuses.

Globalization is another common explanation for human trafficking. Improvement in transportation has made it easy for international commerce as well as human trafficking activities. Increase in global commerce has intensified the awareness of disparities in standards of living and produced a sense of despair for those trapped in impoverished regions. The high demand in sex trade worldwide also takes advantage of the cultural subordination of women in many societies (Miko, 2004). Many societies view women as an economic burden, which may contribute to the selling of daughters and acceptance of brothels as an employment outlet for women.

Masci (2004), on the other hand, argued that although globalization may aid human traffickers it may also improve international collaboration in spotting and fighting against such abuses. Furthermore, globalization may improve the standard of living in developing countries, promote awareness of basic human rights, thus making it harder for predators to recruit unsuspecting victims and helping abolitionist movement to make a difference (Masci, 2004). In this context, what is sorely obvious is a lack of efforts on the part of the governments to combat the issue aggressively. Although poverty may explain much of the initial motivation for outward migration, official corruption has received little attention from researchers. Although research-ers have from time to time pointed out corruption of local law enforcement agencies as a facilitating factor, few have focused on official corruption as a fundamental cause of human trafficking.

Corruption as a Causal Factor

Because poverty is often cited as a key variable in human trafficking, many government agencies and policy makers alike believe that fixing poverty will be the key to eradicating this social ill. Recently some researchers have begun to explore the linkage between official corruption and human trafficking activities. Kevin Bales (2005) has for several years placed corruption as the most significant indicator of human trafficking in countries when compared to other factors such as population or food production. In fact, Bales has pleaded with the academic community to challenge his assertions and findings in the hopes of opening an intellectual debate among critics and supporters (Bales, 2005: 103).

One fundamental challenge in studying corruption is again definitional, as specific socio-cultural and historical configurations give rise to different government operations. There is simply a lack of clear definition of what constitutes corrupt behavior (Morgan, 1998). Instead, different elements have been used to describe corrupt practices. Morgan (1998) stated that the elements of corruption should include nepotism, patronage, personal benefit or diversion of state revenues as well as bribery and cronyism among others. Generally corruption involves irregular conducts by government officials for private gains.

For our purposes, corruption simply means public officials seeking financial reward by either looking the other way or facilitating human trafficking activities. From issuance of travel documents in sending countries to immigration inspection at transit and destination countries, and from control of red light districts to pursuit of trafficking suspects, corrupt officials can be found in many countries and official positions such as consulate officials, border patrol officers, and local law enforcement agencies in districts infested with vice activities. In a report from the *Program against Corruption and Organized Crime in South Eastern Europe* (PACO), the list of corruption officials extends to include intelligence and security services, armed forces, and private businesses such as travel agencies, airlines, and financial institutions (PACO, 2002). The PACO report found that almost all countries in southeastern Europe had corruption problems directly linked to human trafficking and asserted "trafficking cannot take place without the involvement of corrupt officials" (PACO, 2002: 9). These corrupt acts range "from passivity (ignoring or tolerating), or actively participating in or even organizing trafficking in human beings, that is, from a violation of duties, to corruption or organized crime" (PACO, 2002: 7). In many countries, local police officers frequent brothels where trafficked victims are kept as customers (Cockburn, 2003). Visa and immigration officials receive free sexual services in exchange for overlooking fraudulent documents presented by human traffickers (Agbu, 2003). It is no surprise that countries accused of high levels of corruption are also those exerting little effort against human trafficking, whereas states low in official corruption mostly strive to combat human trafficking (PACO, 2002).

Method

In this paper, we operationalize the key variables as follows. The "Corruption Perception Index" (CPI) of 2005 from Transparency International is used as a measure of official corruption. The CPI study has been conducted for many years and expanded to include 146 countries in its most recent analysis. CPI uses a ranking system, in which each country is assigned a number between 1 and 10, 10 being the least corrupt and 1 the worst. The CPI data are typically compiled from 18 different sources and 12 different institutions. The report has emerged as an "index that assembles expert perceptions vis-à-vis corruption" and has become a widely used source for researcher in varying fields (Lambsdorf, 2005: 233).[2]

---

[2] The CPI scoring system measures corruption perceived by business people and country analysts. The score ranges between 10 (highly clean) and 0 (highly corrupt). For instance, according to the 2005 CPI, Iceland, Finland, and New Zealand were ranked the cleanest countries, while Turkmenistan, Bangladesh, and Chad were the most corrupt in the world. Most Scandinavian countries were top ranked. The U.S. was ranked 17th. The CPI gathers data from sources that span the last three years (for the CPI 2005, this includes surveys from 2003, 2004 and 2005). All sources measure the overall extent of corruption (e.g., frequency and/or size of bribes) in the public and political sectors. Data come from surveys as well as experts (risk agencies/country analysts). Each country is also provided a range of possible values (i.e., confidence intervals) that reflect how a country's score may vary. Nominally, the confidence intervals vary about five percent above and below the mean. Additional information on the methodology of the CPI can be found at http://www.transparency.org/policy_and_research/surveys_indices/cpi/2005/methodology.

A country's level of poverty is a somewhat complex construct. Poverty can be measured in different ways. For this study, four different variables are used: (1) per capita income based on a country's gross domestic product (GDP), (2) infant mortality rate under age five, (3) percentage of primary education available to the general population, and (4) life expectancy.[3] Per capita income is the combined output within a country's economy in 1 year divided by the number of population. Infant mortality is measured per 1,000, which indicates the quality of medical care and access to public health services. Percentage of primary education is measured by the number of children that have access to the primary educational systems inside a country. Life expectancy is the average age people in a nation are expected to live. Data are obtained from publicly available sources such as the World Bank, UNICEF, UN Population Division, and the UNESCO Institute for Statistics.

Finally, for the measure of the severity in human trafficking in a nation, we turn to the TIP report. The U.S. State Department has created a three tier system based on the *Trafficking Victims Protection Act of 2000*. The report rates countries based on their compliance with what the State Department calls minimum standards.[4] The first tier applies to countries that have a trafficking problem but are complying with minimum standards to combat trafficking, such as the United Kingdom, Norway, Lithuania, and Czech Republic. The second tier includes countries that are not fully meeting the minimum standards but are making an effort to comply, such as Croatia, Kazakhstan, Paraguay, and Cyprus. A majority of the countries are listed in this tier. The third tier is reserved for countries that are deemed to have a trafficking problem but are doing little or nothing to combat the problem, such as Burma, Jamaica, Saudi Arabia, and Venezuela. A placement on the third tier is subjected to U.S. economic sanctions. The State Department has also created an additional tier in 2003 (reported in 2004) called the Tier 2 "Watch List," that falls between Tiers 2 and 3, for countries that are making significant efforts to move out of the third tier but haven't quite made it. As shown in Table 4.1, some of these

Table 4.1 TIP tier ranking

Year	Tier 1	Tier 2	Tier 3 watch list	Tier 4	Country totals
2001	12	47	–	23	82
2002	18	53	–	19	90
2003	26	75	–	15	116
2004	25	54	42	10	131
2005	24	77	27	14	142

Note: Tier 2 (watch list) in the TIP report is recoded to be Tier 3, and previous Tier 3 to Tier 4

---

[3] Two other variables have also been considered: the official rate of poverty in a nation, and the percentage of child labor. Because of the difficulty in obtaining adequate data from official sources, they are excluded from this study.

[4] The purpose of the minimum standards is to place primary responsibility for the elimination of human trafficking on the government of a country of origin, transit, or destination. Governments in countries with significant trafficking problems are expected to make serious and sustained efforts in curtailing trafficking activities and punishing those involved. Details of the standards and criteria can be found at http://www.state.gov/g/tip/rls/tiprpt/2005/46770.htm.

countries include China, India, Sierra Leone, and Cameroon. Dummy variables are created for each of the four tiers in the TIP report to indicate whether or not a nation is in a certain tier.

The TIP tier system has come under much international criticism because it is often used by politicians in the U.S. to tie foreign aid to efforts in combating human trafficking. Many countries' placements in the tier system are also rather perplexing. Countries such as Greece and Turkey, both U.S. allies and NATO members, have failed to get better than a Tier 2 rating over the past 5 years (Miko, 2004). On the other hand, the U.S. does not subject itself to this rating system. Countries that many researchers and NGOs believe as having serious trafficking problems, such as Russia and Pakistan, receive higher ratings for political reasons (Miko, 2004).

Despite its many shortcomings, the TIP report provides one source of information as well as a database at least for a rudimentary analysis of human trafficking problems in the world. The TIP report attempts to examine not only the levels of human trafficking in different countries but more importantly governmental and legislative efforts to combat the problem. Countries with well articulated legislations and bone fide enforcement agencies in pursuing human traffickers usually are placed in Tier 1. Furthermore, the TIP Report and the data presented in it have improved over the years. The report also helps the State Department determine the level of financial aid to foreign countries to curb human trafficking. For example, "in 2003 the U.S. Government alone supported 190 anti-trafficking programs in 92 countries, totaling U.S. $72 million" and the amount has been increasing over the years (Laczko, 2005: 6).

Even though the United States is not included in the TIP Report, the State and Justice Departments are fully aware of the trafficking dilemma within the U.S. borders. Numerous cases of human trafficking have been brought to the forefront by law enforcement raids and prosecutions that have been widely reported in the news. These cases have been found across the U.S. in various forms. Sweatshop labor and sexual slavery have been a part of the U.S. reality for years. Very much like trafficking around the globe, the number of the victims and the extent of trafficking organizations are sketchy.

Analysis and Findings

The purpose of this study is to evaluate the relative impact of poverty and official corruption on human trafficking. We hypothesize corruption to be just as a significant factor in human trafficking as poverty. We begin our analysis by comparing the means of all key predictor variables in each TIP tier, as shown in the tables, to provide descriptive statistics on how these variables may relate to human trafficking. Table 4.2 reveals that per capita income appears to be a strong predictor of human trafficking as countries in Tier 1 are almost $18,000 above the next category, which is Tier 4. Countries in Tier 2 rank third even though Switzerland has a per capita income of $49,300, which shows in a wider spread of standard deviations. Tier 3 (the watch list) category has the lowest per capita income, but Tier 4 categories have greater wealth but the worst trafficking situations. There are two ways to interpret these mixed patterns in Table 4.2. First, per capital income predicts well only within Tier 1, but not the other three categories. One would expect the lower the per capita income, the worse the trafficking problems as more prospective victims are available for exploitation. Countries in Tier 4 should have the lowest average per capita income. But this is not the case. A review of the Tier 4 countries (Tier 3 in the TIP report) reveals that several wealthy nations also have the worst trafficking problems. These countries include Qatar, United Arab Emirates, Saudi Arabia, and Kuwait. Because of these petroleum-rich countries, per capita income is artificially inflated. Second, as the TIP tier system attempts to point out, despite the presence of serious human trafficking activities, these countries are in this tier because there have been few governmental and legislative efforts to pursue the perpetrators and to protect the victims. In either case, per capita income does not appear to be a good predictor for the severity of human trafficking problems.

A comparison of the average life expectancy of the countries reveals a similar pattern as per capita income above, as shown in Table 4.3. Tier 1 produces the highest life expectancy on average, and Tier 4, considered the worst in the fight against human trafficking, has the second highest life expectancy. Tier 2 countries rank the third, which leaves Tier 3 "watch list" countries the last in life expectancy. These last two tiers also have the largest variations, as shown in their standard deviations, among the countries. For instance, Switzerland with a life expectancy of 81 skews the mean in a positive direction in Tier 2. Without Switzerland, both Tier 2 and Tier 3 would have been very similar. In any case, life expectancy (which is related to a nation's level of poverty) does not appear to be a good predictor of the trafficking problem in a nation.

Infant mortality also follows similar patterns presented thus far. Tier 1 holds the best infant mortality rate of the four categories, and Tier 3 "watch list" has the

Table 4.2 Per capita income by TIP ranking

Tier ranking	Mean	Std. deviation	N
1	$26,583.42	17274.16	24
2	$4,901.25	8767.23	76
3	$2,951.63	4367.01	27
4	$8,699.17	12139.45	12
Total	$8,594.11	13299.16	139

Table 4.3 Life expectancy by TIP ranking

Tier ranking	Mean	Std. deviation	N
1	78.2	2.33	20
2	64.4	12.34	75
3	61.9	12.20	27
4	68.1	8.35	14
Total	66.35	12.11	136

worst. However, both Tier 2 and Tier 3 categories have wide variations in their mortality rates as shown in standard deviations. Tier 4 countries, with their worst record in fighting against human trafficking, has a better infant survival rate than their counterparts in Tier 2 or Tier 3, which indicates an overall better socio-economic environment or greater access to healthcare services. Infant mortality rate as a measure of poverty or socio-economic development is again not a good predictor of how a nation deals with the human trafficking problem (Table 4.4) .

Access to primary education in general population is often considered a good measure of a nation's financial strength and economic development. As shown in Table 4.5, although Tier 1 countries stand clearly ahead, countries in the other three categories are close to one another. Since most countries in Tier 1 are in the developed world, access to primary education is readily available to their populations at the rate of 95.6% on average. However, there appears to be few differences among countries of the other three tiers. Regardless of their rank in the tier system, most countries these three tiers are quite similar in their provision of primary education to the general population. In other words, access to primary education, with the exception of Tier 1 countries, appears to have little impact on a nation's effort to fight against human trafficking.

Finally, corruption (i.e., the CPI index) is measured on a ten point system, 1 being the most corrupt and 10 the least. Similar to findings on the effect of per capita income, Tier 1 countries stand well ahead of those in the other three categories, as shown in Table 4.6. Moreover, there appears to be a consistent descending order among Tiers 1, 2, and 4 in corruption scores. Tier 3 (the watch list) countries have the worst corruption scores on average. Without the "watch list" countries, which are ambiguously constructed, the descent among Tiers 1–4 is relatively clean and consistent. When Tiers 3 and 4 are combined, their mean is 3.23, again lower than the average of Tier 2.

Table 4.4 Infant mortality (under five) averages by TIP ranking

Tier ranking	Mean	Std. deviation	N
1	6.25	3.64	20
2	71.75	72.83	75
3	82.67	76.87	27
4	51.14	45.04	14
Total	62.16	69.67	136

Table 4.5 Percentage of primary education available to population by TIP ranking

Tier ranking	Mean	Std. deviation	N
1	95.60	9.42	10
2	82.81	16.94	69
3	83.23	15.98	26
4	82.50	12.50	12
Total	83.97	16.06	117

Table 4.6 Official corruption averages by TIP ranking

Tier ranking	Mean	Std. deviation	N
1	6.80	2.24	24
2	3.65	1.87	66
3	3.10	1.01	25
4	3.37	1.42	11
Total	4.12	2.20	126

All the above predictor variables appear to possess some explanatory power. On the other hand, official corruption seems to have more consistency in its prediction on a nation's standing in the TIP report than all other variables. Although these mean comparisons produce some general patterns about a nation's fight against human trafficking, a lot more can be done to further our understanding on how a government's transparency in its operations may affect the trafficking problem in a predictable manner.

After observing the general patterns from the above descriptive statistics, we proceed to conduct correlation and regression analyses with these variables. Table 4.7 presents the Pearson correlation coefficients of these six variables. All predictor variables have coefficients that reach significant levels, except for percent of population with access to primary education. Of all variables, corruption ($r = -0.44$) and per capita income ($r = -0.41$) seem to be most correlated with the TIP ranking. In other words, the greater the transparency in government operations (less corruption) the lower the rank in the TIP tier system (less problem with human trafficking); and the greater the per capita income the lower the TIP rank.

We further examine the relative beta weights of these predictor variables in a regression model, as shown in Table 4.8. Although the regression model contains

Table 4.7 Pearson correlation of all key variables

	1	2	3	4	5	6
Tier ranking	–	0.230**	−0.435***	−0.414***	−0.286	−0.081
Infant mortality	0.230**	–	−0.446***	−0.448***	−0.928***	−0.541***
Corruption	−0.435***	−0.446***	–	0.845***	0.534**	0.235**
Per capita income	−0.414***	−0.448***	0.845***	–	0.552***	0.252***
Life expectancy	−0.286***	−0.928***	0.534***	0.552***	–***	0.591***
Primary education	−0.081	−0.541***	0.235**	0.252***	0.591	–

Note: **$p < 0.01$; ***$p < 0.001$

Table 4.8 Regression of predictor variables on TIP tier ranking

Predictor variables	B	Std. error	Beta	t	Sig.
(Constant)	3.62	1.37		2.64	0.01
Infant mortality	0.00	0.00	−0.09	−0.45	0.66
Corruption	−0.13	0.07	−0.28	−1.92	0.06
Per capita income	0.00	0.00	−0.13	−0.89	0.38
Life expectancy	−0.02	0.02	−0.20	−0.86	0.39
Primary education	0.01	0.01	0.09	0.90	0.37

Model statistics: $R^2 = 0.21$; adj. $R^2 = 0.18$

only five independent variables, it produces a moderate R^2 value that explains about 18% of variance in the TIP tier ranking system. Although most predictor variables are significantly correlated with the dependent variable (i.e., TIP ranking) in the anticipated direction, only corruption carries a beta weight that approaches statistical significance ($p = 0.056$) in the regression model. All other predictor variables fail to achieve statistical significance in their relative impact on the dependent variable.

Discussion

Human trafficking is a complex social issue that ranges from forced labor to sexual exploitation. Many causal factors interplay to influence this modern day slavery. Trafficking and victimization trends often vary from region to region, which makes it hard to develop coherent and consistent causal patterns. The fact that most victims come from less developed countries has led researchers to focus almost exclusively on the role of poverty in the trafficking business. The poverty theory makes sense, because the lack of economic opportunities most often provides the initial impetus for prospective victims to fall prey to human traffickers. However, in order for human traffickers to mount successful and continuous operations, a regulatory or socio-legal environment conducive to the trafficking trade and related vice industry and labor market is required. This study attempts to raise the profile of this regulatory environment in search of not just factors contributing to the onset of human trafficking but more importantly what serves to sustain and perpetuate this sinister enterprise. Both the descriptive and multivariate statistical analyses, as shown in the above tables, seem to suggest that although certain macro-level variables possess some explanatory power, corruption is probably the most important factor in explaining human trafficking.

Most of what we know thus far is heavily influenced by reports and documents produced by advocacy groups, non-government organizations, and government agencies. Unfortunately, most current research on human trafficking has made little progress over the years because of the secrecy of the business and the political sensitivity of the topic. Most published studies thus far have largely relied on anecdotal stories or sporadic personal observations. Weitzer (2005) argues that the growing moral panic over prostitution and sex trafficking is most responsible for the current escalation of political discussion on human trafficking in western countries. The urgency in drafting U.S. government policy and funding anti-trafficking organizations worldwide has largely occurred with little empirical verification.

This study makes an exploratory attempt to include official corruption in the main discourse on human trafficking, and to test the hypothesis that corruption may be a stronger predictor than other macro level factors in explaining the presence of trafficking activities in certain countries. This hypothesis is developed in contrast to the popular assumption that human trafficking is mainly a product of poverty and lack of economic opportunities. While there is no denying that women and children, allured by financial incentives, often fall prey to traffickers, the socio-cultural circumstances that permit and sustain not only the initial deception and subsequent enslavement of prospective victims may have more to do with governmental ineptitude or downright corruption. To corroborate the corruption hypothesis, authors of this study have also conducted field observations in the red light district in Tijuana, Mexico. These field activities, which are reported elsewhere, strongly suggest that official corruption may indeed sustain and even encourage the trafficking business.

There are several limitations in this study. First of all, we rely on official data. Second, some of the variables can only be considered proxy measures of what might have been the situation. For instance, the corruption perception index is not a direct measure of corruption per se; rather, it is merely a measure of perception. The TIP tier system is also controversial. It is not a direct measure of the level of the human trafficking activities in a country. The ranking system reflects the U.S. government's perception of a foreign government's effort in cracking down on transnational human trafficking activities. Moreover, much of the U.S. State Department's TIP ranking is influenced by reports from NGOs and other activist groups.

A direct measure of the severity of human trafficking activities is impossible, which is not surprising. Proxy measures therefore need to be developed to gauge the problem. Because measures directly measuring clandestine criminal activities are not likely, we look for indicators and apply those for analysis purposes. A lack of government effort to combat human trafficking is highly correlated with the severity of trafficking problems, at least conceptually. Countries that make the least effort to fight human trafficking also tend to be those with high levels of official corruption. Short of empirical evidence to suggest otherwise, our only practical solution is to put forth a proxy measure to test our hypothesis. In other words, corruption impedes a nation's political will and effort to combat sex trafficking.

Because research on human trafficking is still in its infancy, much more work is needed to develop and test theoretical arguments. We do not dismiss poverty as an important variable, but call for increased attention to official corruption as a valid predictor of human trafficking activities. A lack of government transparency, official ineptitude, and collusion are likely to facilitate and enable human trafficking to a greater extent than poverty-related factors. The policy implication in this study is clear – any effort to curb human trafficking must address its symbiotic relationship with the regulatory environment.

Appendix: 2005 State Department Tip Report Tier Placements

Tier 1

Australia, Denmark, Luxembourg, Poland, Austria, France, Morocco, Portugal, Belgium, Germany, Nepal, South Korea, Canada, Hong Kong, The Netherlands, Spain, Colombia, Italy, New Zealand, Sweden, Czech Republic, Lithuania, Norway, United Kingdom.

Tier 2

Afghanistan, Egypt, Laos, Senegal, Albania, El Salvador, Latvia, Serbia-Montenegro, Algeria, Equatorial Guinea, Lebanon, Singapore, Angola, Estonia, Libya, Slovenia, Argentina, Ethiopia, Macedonia, Sri Lanka, Bangladesh, Finland, Madagascar, Switzerland, Belarus, Gabon, Malawi, Syria, Bosnia/Herzegovina, Georgia, Malaysia, Taiwan, Brazil, Ghana, Mali, Tajikistan, Bulgaria, Guatemala, Mauritania, Tanzania, Burkina Faso, Guyana, Moldova, Thailand, Burundi, Honduras, Mongolia, Turkey, Chad, Hungary, Mozambique, Uganda, Chile, Indonesia, Nigeria, Uruguay, Congo, Iran, Oman, Vietnam, Costa Rica, Israel, Pakistan, Yemen, Cote D'ivoire, Japan, Panama, Zambia, Croatia, Kazakhstan, Paraguay, Cyprus, Kenya, Peru, East Timor, Kyrgyz Republic, Romania.

Tier 2 Watch List (Recoded to be Tier 3 in This Paper)

Armenia, Dominican Republic, Mexico, Slovak Republic, Azerbaijan, Gambia, Nicaragua, South Africa, Bahrain, Greece, Niger, Suriname, Belize, Guinea, Philippines, Ukraine, Benin, Haiti, Russia, Uzbekistan, Cameroon, India, Rwanda, Zimbabwe, China, Mauritius, Sierra Leone.

Tier 3 (Recoded to be Tier 4 in This Paper)

Bolivia, Ecuador, Qatar, United Arab Emirates, Burma, Jamaica, Saudi Arabia, Venezuela, Cambodia, Kuwait, Sudan, Cuba, North Korea, Togo.

Acknowledgments The authors would like to acknowledge the helpful comments from the anonymous reviewer and the contribution by Alex Franco to the earlier conceptualization of the thesis in this paper.

References

Agbu, O. (2003). Corruption and human trafficking: The Nigerian case. *West Africa Review, 4*(1). Accessed on February 26, 2006. <http://www.westafricareview.com/vol4.1/agbu.html>.

Associated Press. (2005). Major child prostitution rings broken: 31 indicted. *San Diego Union Tribune*, December 17. sec. A7.

Bales, K. (2005). *Understanding global slavery: A reader*. Berkeley and Los Angeles, CA: University of California.

Belser, P., Cock, M. de, & Mehranl, F. (2005). *ILO minimum estimate of forced labour in the world*. Geneva: ILO.

Cockburn, A. (2003). 21st century slaves. *National Geographic Magazine* (September). Accessed on February 26, 2006. <http://magma.nationalgeographic.com/ngm/0309/feature1/>.

Feingold, D. A. (2005). Think again: Human trafficking. *Foreign Policy Issue, 150*, 26–32.

Gozdziak, E. M., & Collett, E. A. (2005). Research on human trafficking in North America: A review of literature. *International Organization for Migration, 43*(1/2), 99–128.

International Labor Organization. (2005). A global alliance against forced labor. *International Labor Conference 93rd session*. Accessed on February 26, 2006. <http://www.ilo.org/dyn/declaris/DECLARATIONWEB.DOWNLOAD_BLOB?VarDocumentID = 5059>.

Laczko, F. (2005). Data and research on human trafficking: A global survey. *International Organization for Migration, 43*(1/2), 1–339.

Lambsdorf, J. G. (2005). *Corruption perceptions index. Global corruption reports*. London. Ann Arbor, MI: Transparency International.

Masci, D. (2004). Human trafficking and slavery: Are the world's nations doing enough to stamp it out? *CQ Researcher, 14*(12), 273–296.

Miko, F. (2004). Trafficking in women and children: The U.S. and international response. *Foreign Affairs, Defense, and Trade Division. Congressional Research Service Report*, 98–649C.

Morgan, A. L. (1998). Corruption: Causes, consequences, and policy implications. A literature review. *The Asia Foundation Working Paper Series. Working Paper # 9*. Accessed on February 26, 2006. <http://www.asiafoundation.org/Publications/workingpapers.html>.

Outshoorn, J. (2005). The political debates on prostitution and trafficking of women, social politics. *International Studies in Gender, State and Society, 12*(1), 141–155.

PACO. (2002). *Trafficking in human beings and corruption* (report on the regional seminar). Program against corruption and organised crime in South Eastern Europe. Economic Crime Division. Portoroz, Slovenia.

Rathgeber, C. (2002). The victimization of women through human trafficking –An aftermath of war? *European Journal of Crime, Criminal Law and Criminal Justice, 10*(2/3), 152–163.

Siobhan, M. (2006). Sinister industry. *ABA Journal, 92*(3), 59–60.

United Nations. (2000). Protocol to prevent, suppress and punish trafficking in persons, especially women and children, supplementing the United Nations convention against transnational organized crime. Palermo, Italy. Accessed on February 26, 2006. <http://www.uncjin.org/Documents/Conventions/dcatoc/final_documents_2/convention_%20traff_eng.pdf>.

U.S. Department of State. (2005). Trafficking in persons report. *Office to monitor and combat trafficking in persons*. Accessed on February 26, 2006. <http://www.state.gov/g/tip/rls/tiprpt/2005/>.

Weitzer, R. (2005). The growing moral panic over prostitution and sex trafficking. *The Criminologist, 30*(5), 1–5.

Chapter 5
New Players in an Old Game: The Sex Market in Italy

Stefano Becucci

Through the analysis of data provided by the National Statistics Institute (Istat) and court records this essay examines the sex trade in Italy, with particular attention devoted to street prostitution. Changes recorded over the past decade will be shown to correspond to the progressive replacement of Italian criminal groups with foreign groups. What is more, these foreign associations appear to adopt different organizational models in recruiting immigrant women destined for the sex trade. Moreover it focuses attention on the types of relationships between Italian mafia-type associations and foreign exploiters. Considering the existence of indigenous mafia organizations rooted in the territory, at issue here is whether or not groups of foreign exploiters operate largely autonomously in managing street prostitution or, on the other hand, are affected by the "cumbersome" presence of local mafia associations.

Market Characteristics

Similar to the situation occurring in other countries of the European Union, street prostitution in Italy has undergone profound transformations over the past few years (Omicron, 2001: 38). The progressive disappearance of Italian prostitutes has been witnessed since the end of the 1980s. Increasing dangers in street work, and the absence of any generational replacement for local women, have made room for a new female population arriving from various other countries. Italian women have continued to ply their trade in apartments, considered safer locations, frequently relying on regular customers they had become familiar with during their previous street activity (Corso & Landi, 1998: 211).

The arrival of foreign women has been marked over time by a series of *waves*. The first, between 1989 and 1990, was mostly made up of women from Eastern Europe, particularly Bosnian, Slovenian, and Romanian, and began after the fall of the Berlin Wall and the subsequent conflict between Croatia and Serbia. The second, in 1991–1992, saw the considerable presence of African women, largely from Nigeria, who arrived directly from Lagos and Benin City or via Amsterdam and Rotterdam. Young Albanian women composed the third major wave between 1993 and 1996. More recently, women have started arriving from countries of the former

D. Siegel and H. Nelen (eds.), *Organized Crime: Culture, Markets and Policies.*
© Springer 2008

Soviet Union, predominantly Ukraine, Moldavia, Bulgaria, and Russia (Carchedi, 2000: 121–124; 2004: 77–105).

The sizeable presence of foreign women has led to a drop in prices. A foreign prostitute working on the street gets between 15 and 30 euros per "trick," a price only slightly higher than what Italian women earned 15 years ago[1] (Corso & Landi, 1998: 213; Dal Lago & Quadrelli, 2003: 222; Tribunale Genova, 2001: 33). In the past, the *rules* of street prostitution seemed to have been markedly different. One streetwalker states, "[during the period of] us professionals, you could tell right away if a girl was undercutting prices; the talking drums on the street worked fast, the customer himself told you. A sort of self-regulation started up: a group immediately went to whoever had been trying to sell herself below rates and, with more or less gentle methods, convinced her to keep the same prices" (Corso & Landi, 1998: 214).[2]

Another major change has been prostitution's shift from stationary to mobile. While previously Italian women usually operated in established spots and relied on a regular clientele, today continuous movement is common from one city to another along with fleeting, hurried relationships with customers. This frequent transfer of prostitutes helps revive demand and at the same time fills the functional needs of the organizations managing this market. By frequently changing location, exploiters keep their women from establishing any strong relationships with customers that, subsequently, might help free them from the system of exploitation.

The sex market is structured in various forms: from street prostitution to "night clubs," from upscale prostitutes – call-girls who work with a select group of clients – to individual prostitutes who place announcements in newspapers, and the growing number of internet sites and television channels that offer various types of sexual services (Taylor & Jamieson, 1999: 265). Some estimates place the number of prostitutes present in the Italian market at approximately 50,000; half of these perform their activities on the street, while the rest work mainly in apartments and nighttime locales (Monzini, 2002: 13). The greatest concentrations of sex workers are in Northern Italy, particularly the cities of Milan, Venice, Bologna, Rimini, and Torino: either in especially populous areas or places like Rimini on the Adriatic Sea that attract large masses of young people in the summer. In Central Italy, the cities most involved with this phenomenon are Florence and Prato; the most important areas in the South are represented by the cities of Naples and Caserta (Carchedi, 2000: 100–162).

---

[1] While Italian street prostitutes earned 50,000 lire per service in the early 1980s, today foreign prostitutes charge around 30 euros (1 euro equals slightly less than 2,000 lire). Although the "cost" of services in the street sex trade has remained nominally almost the same, there has nonetheless been a significant decrease when taking into account the lower buying power of the currency compared to several decades ago.

[2] Some distortion should be taken into account in the stories from women who worked as streetwalkers in the past, resulting from overly positive recollections of their experiences in the sex trade. There is reason to believe that violence by pimps was also widespread during the 1970s, though probably in a less organized and systematic form than today.

Italians and Foreigners Involved in the Exploitation of Prostitution

Before explaining the different organizational models of the criminal actors, we should examine an analysis of judicial statistics related to the crime of exploiting and aiding and abetting of prostitution (see Tables 5.1–5.7) Even considering potential distortions tied to specific methods of *producing* this statistical data, foreigners still assume increasing importance over time and reach almost half the total.[3]

Several interesting aspects emerge through analysis of the foreign communities most involved in this type of crime; they originate from Albania, some Eastern European countries, and Nigeria. Albanian exploiters show great involvement over the period under examination, much higher than any other foreign group. The number of people from Nigeria is also increasing and has reached second place over the past few years among the ten nationalities most involved in this activity. The latter make up almost the total of foreigners with penal actions brought against them: percentages run from 78% in 2003 to 98% in 1995. Though a relative expansion of national groups has been witnessed over the past few years, the overall criminal phenomenon remains firmly concentrated among the top three or four nationalities. Finally, in 2003, the last year available, the growing presence was recorded of sex exploiters from Romania along with a noticeable decline in Albanians compared to previous years.

---

[3] Regarding potential distortion of this statistical data, we refer to two sorts of problems. The first pertains to the system of codifying judicial statistics at Istat (the National Statistics Institute). These statistics make no reference to data related to the crime of trafficking women for the purpose of sexual exploitation, but only to the most probable consequences such as sexual exploitation. In addition, the same penal category includes both exploitation as well as aiding and abetting prostitution, two crimes that may take on quite different meanings depending on specific circumstances. The second problem regards the potential existence of forms of discrimination carried out by law enforcement under pressure from the media and politicians looking to exploit forms of social alarm in the population for electoral ends. Although Italy may have for some time experienced growing forms of intolerance, xenophobia and racism that may also have at least somewhat influenced law enforcement operations and given the increase in reports to the police by the general public, the fact nonetheless remains that these judicial statistics reflect powerfully differentiated criminal involvement, from a quantitative as well as a qualitative perspective (related to the *modus operandi* of the criminal groups) with respect to the various foreign associations involved in sexual exploitation. Although detailed analysis of this problem would require more specific study beyond the scope and limited space of this article, if the main factor at the basis of the involvement of certain foreign groups in managing prostitution were tied to discriminatory acts by law enforcement, we would have to attempt to explain why, for example, immigrants from China and Asia in general comprise an irrelevant statistical presence in this crime, even though they experience forms of discrimination, intolerance and separation from the social context on a par and perhaps even more so than other immigrant groups. As shown below, the lower presence of prostitution in the Chinese community seems to be tied to the existence of informal social controls within the established community itself.

Table 5.1 Italians and foreigners reported for exploitation and aiding and abetting of prostitution by the Judicial Authorities (1995–1998)

	1995	%	1996	%	1997	%	1998	%
Foreigners	476	39	550	41	562	46	562	47
Total Italy	1,235	100	1,329	100	1,216	100	1,191	100

Source: personal calculations from Istat data

Table 5.2 Italians and foreigners reported for exploitation and aiding and abetting of prostitution by the Judicial Authorities (1999–2003)

	1999	%	2000	%	2001	%	2002	%	2003	%
Foreigners	658	53	458	47	622	48	575	45	584	48
Total Italy	1,241	100	981	100	1,283	100	1,289	100	1,205	100

Source: personal calculations from Istat data

Table 5.3 Top ten foreign nationalities most involved in the crime of exploitation and aiding and abetting of prostitution (1995–1996)

Nationality	1995	%	Nationality	1996	%
Albania	266	56	Albania	235	43
Former Yugoslavia	121	25	Former Yugoslavia	146	27
Nigeria	21	4	Nigeria	31	6
Tunisia	15	3	Brazil	16	3
Former Soviet Union	10	2	Tunisia	13	2
Romania	9	2	Morocco	13	2
Ghana	8	2	Ghana	9	2
Austria	5	1	Former Soviet Union	6	1
Brazil	5	1	Former Czechoslovakia	5	1
Colombia	5	1	Poland	5	1
Top ten total	465	98	Top ten total	479	87
Total foreigners	476	100	Total foreigners	550	100

Source: personal calculations from Istat data

Table 5.4 Top ten foreign nationalities most involved in the crime of exploitation and aiding and abetting of prostitution (1997–1998)

Nationality	1997	%	Nationality	1998	%
Albania	263	47	Albania	234	42
Former Yugoslavia	90	16	Former Yugoslavia	69	12
Ghana	67	12	Nigeria	66	12
Nigeria	28	5	Romania	37	7
Romania	19	3	Former Soviet Union	27	5
Tunisia	12	2	Brazil	25	4
Brazil	9	2	Colombia	11	2
Morocco	7	1	Bulgaria	9	2
Austria	6	1	Former Czechoslovakia	8	1
Chile	6	1	Tunisia	8	1
Total top ten	507	90	Total top ten	494	88
Total foreigners	562	100	Total foreigners	562	100

Source: personal calculations from Istat data

Table 5.5 Top ten foreign nationalities most involved in the crime of exploitation and aiding and abetting of prostitution (1999–2000)

Nationality	1999	%	Nationality	2000	%
Albania	296	45	Albania	224	49
Nigeria	73	11	Nigeria	35	8
Romania	59	9	Serbia-Montenegro	25	5
Former Yugoslavia	54	8	Romania	24	5
Brazil	28	4	Colombia	14	3
Former Soviet Union	24	4	Ukraine	10	2
Colombia	21	3	China	9	2
Tunisia	9	1	Bulgaria	8	2
Morocco	8	1	Dominican Republic	7	2
Former Czechoslovakia	7	1	Tunisia	7	2
Top ten total	579	88	Top ten total	363	79
Total foreigners	658	100	Total foreigners	458	100

Source: personal calculations from Istat data

Table 5.6 Top ten foreign nationalities most involved in the crime exploiting and aiding and abetting prostitution (2001–2002)

Nationality	2001	%	Nationality	2002	%
Albania	276	44	Albania	174	30
Nigeria	50	8	Nigeria	66	11
Colombia	45	7	Romania	61	11
Romania	35	6	Colombia	37	6
Ukraine	27	4	Brazil	30	5
Serbia-Montenegro	24	4	Serbia-Montenegro	29	5
Brazil	21	3	Bulgaria	19	3
Morocco	14	2	Dominican Republic	16	3
Dominican Republic	13	2	Ukraine	16	3
Russian Federation	12	2	Russian Federation	14	2
Top ten total	529	85	Top ten total	462	80
Total foreigners	622	100	Total foreigners	575	100

Source: personal calculations from Istat data

Table 5.7 Top ten foreign nationalities most involved in the crime of exploiting and aiding and abetting prostitution (2003)

Nationality	2003	%
Albania	162	28
Romania	113	19
Nigeria	65	11
Colombia	25	4
Dominican Republic	20	3
Brazil	19	3
China	17	3
Bulgaria	13	2
Tunisia	12	2
Ukraine	10	2
Top ten total	456	78
Total foreigners	584	100

Source: personal calculations from Istat data

Foreign Criminal Actors: Organizational Models

The characteristics of the various criminal groups are different enough to help draw the outline of several reference models, each made up of specific organizational modalities.

The Albanian Model

Albanian groups are usually composed of a small number of individuals; each of these "manages" two or three prostitutes. Members of the organization frequently come from the same cities or villages in Albania and, in some cases, share blood ties: brothers, cousins, and sometimes uncles and nephews participate, along with other compatriots, in the exploitation of prostitution. Certain *rules* govern the management of women being exploited. First of all, if the individual exploiter has brought one or more women to Italy, then he has more or less exclusive rights over her. Her pimp's agreement is required in the case of decisions that could put the woman's life at risk. As one victim testifies:

The reason I was worried [Author's Note: *the girl had discovered she was pregnant*] was the fact that although, on the one hand, I had always taken every precaution to avoid disease or undesired pregnancy on the "job," on the other, whenever my keepers wanted to let off steam with me sexually, they avoided using prophylactics. So I was afraid of being pregnant again and this appeared to trouble the men quite a bit. After some discussion, one of the most powerful members of the group in Bergamo, B., intervened and asked A. [Author's Note: *the cousin and pimp of the girl he had brought to Italy*] for authorization to resolve "my" problem using the same technique on me he had used on another girl, according to him with effective results. A. told him to do whatever he wanted as long as it took care of the situation. B. said it would be necessary to hold me down... (Tribunale Genova, 2001: 95).

Even in the presence of a potentially exclusive relationship between victim and her "owner" – reinforced by sentimental ties frequently linking female countrymen to their exploiters – the group logic prevails in many situations. For example, if a pimp is absent for a set period of time, another is chosen who must keep "his" women under control. When one member is arrested by law enforcement, the others decide to use a percentage of their profits for his legal assistance and maintenance in jail. The entire group altogether decides to move if police controls tighten or they hear profits are higher in other Italian cities.

As much as this may be similar to the typical model of the pimp who creates an ambiguous relationship with his prostitute/lover, the Albanian exploiters' *modus operandi* shows several specific elements that distance them from this overall picture. These individuals practice extremely brutal acts of violence. Girls are beaten until they collapse, or dangled from balconies with threats to drop them if they do not bow to the will of their "bosses." If they come up short of the pre-set daily amount of money, they are systematically beaten and tortured.

These exploiters constantly turn to heinous practices of violence, often apparently unmotivated and gratuitous. This violence has the specific purpose of instilling within the victim the rules governing her relationship with the exploiter. The main goal of sexual violence itself is to subdue the victim and render her tame and obedient. The synergies created between the exploiters and the methodical use of violence keep their victims under constant threat. The latter have no margin of autonomy. They are continuously under surveillance when they work on the street and pimps count the unused prophylactics at the end of their shift in order to verify the number of customers. In the end, in order to block any relationships with customers that could give room for escape, women are called every 5 min on their cell phones, forcing them to show themselves on their assigned "spot" on the street.

The criminal network of Albanian exploiters seems to be characterized by the lack of major transnational ties between those recruiting the women and those in Italy who force them into prostitution. Although it is difficult to establish a clear outline of connections existing between traffickers and exploiters from Eastern Europe operating in Italy, several elements emerge from an analysis of judicial reports that corroborate this interpretation. Casting doubt on plausible ties with criminal structures existing in their country of origin, Albanian exploiters usually rely on traffickers located at various junctions along the route between one country and the other, in places they know they can find women to put into prostitution. They personally "buy them" in Albania, or bordering countries like Serbia and Montenegro, and then bring them to Italy.

Romanian, Moldavian, and Ukrainian women refer to having been "bought and sold" many times before arriving in Italy, at times having up to ten "bosses." This probably indicates a certain fragmentation among criminal actors who prefer to exploit the women for a short time, sometimes even just a few days and, immediately afterward, hand them over to new exploiters. This sudden handoff from one boss to another could be dictated by the fact they have constructed no reference network capable of managing the entire process of exploitation, from the recruitment phase up to final placement in the Italian market.

The choice aimed at maximizing "individual" profits – instead of those of the whole group of actors involved in the market, who would gain greater advantage if they were able to manage the entire process – probably shows a certain fragmentation among criminal groups involved in the exploitation of prostitution. Though various types of segmentation mark the system of recruitment and exploitation, victims' accounts reveal the existence of junctions along established routes in Albania, Serbia, and Montenegro, where exploiters likely converge in search of women destined for prostitution (Tribunale Lecce, 2000a: 2; 2000b: 3).

Finally, major changes should be noted in the realm of street prostitution over the past few years. Some associations and operators, occupied in rescuing women who have suffered under regimes of slavery and sexual exploitation, claim that, since early 2000, street prostitution has been gradually shifting to indoor locations, such as apartments or night clubs, that guarantee greater privacy for the customer and reduce the risk for the women, as well as their exploiters, of being identified by law enforcement (Bufo, 2005: 15). This is due, firstly, to the converging effects of various factors such as greater police controls on street prostitution, the increase in the risk of deportation for undocumented immigrants as foreseen in 2002 by the latest law against clandestine immigration, and, secondly, to the legislation proposed in December of the same year that would ban street prostitution in exchange for the possibility of practicing it indoors, in well-defined, restricted urban areas. As if the organizations involved in this market had decided to take these dreaded legislative changes into account, some signs indicate the existence of a growing connection between indoor and street prostitution, spheres that a short while ago were considered separate and non-communicating because of their target customers, price ranges, and the sexual services offered (Donadel & Martini, 2005: 35). Recently, a trend has been noted toward a decrease in violence by Albanian exploiters and a relaxation of direct control over their women, in various cases substituted by prostitutes/lovers who stand in for them and assume the concrete risk of managing a group of "colleagues" (Carchedi, 2003: 133–134).

The Nigerian Model

If exploiters from Albania and Eastern Europe in general, are characterized by being mainly involved in managing prostitution, Nigerian organizations, on the other hand, control recruitment as well as the subsequent introduction into the Italian sex market.

Major figures supervising recruitment in the country of origin are the so-called *madam*, a sort of priestess, and the *sponsor*. The first has the job of using magic rituals from *juju* (a local variation of traditional *voodoo*) to bind girls to the agreements stipulated by the organization; the second advances travel costs, equal to around 6,000$ a person. On arriving in Italy, girls are placed under the control of other madams (different from the previous since they usually do not perform magic rituals). In some cases, there is also the figure of the *controller*, frequently an ex-prostitute, who has the assignment of watching over the women and guaranteeing the "contract" agreed to on departure with the criminal organization is respected (Massari, 2003: 316–320; Monzini, 2002: 71).

The definition of an "agreement," written or verbal, obliges the victim to respect certain rules of conduct. Nevertheless, working conditions and promised earning potential rarely correspond to reality. Initially, the traffickers cover transportation expenses and promise the women this debt will be paid off after just a few months work. In reality, on arriving in Italy, girls discover most of their earnings are confiscated by their criminal contacts and a long time will pass before they are out of debt. Alongside this, other expenses such as rent, the cost of prophylactics and other necessities related to street work are "loaded" onto the girls directly by their exploiters (Questura Udine, 2000: 19). Victims must prostitute themselves for several years until they have paid off the debt, usually around 40,000–50,000 euros.[4]

---

[4] Interview with a functionary from the International Organization for Migration, Rome, 12 May 2003.

Taking advantage of animist beliefs that still persist in some areas of Nigeria, these criminal associations use magic rituals to force the girls into an *ironclad contract* that, though bloodless, nevertheless proves to be extremely binding. The *ritual* calls for the girls to hand over some of their personal effects, along with bits of hair and fingernails that are then mixed with powdered animal bone and finally rolled into a scroll bearing the name of the girl and kept by the madam. The ritual has the power of inspiring real terror in victims who believe they will be seized by madness or sudden death if they fall short of the promises made.

Enslavement strategies adopted by exploiters from Eastern Europe[5] are based on the ruthless, systematic use of violence, while those used by Nigerian traffickers are mostly identified by recourse to forms of psychological suggestion, utilizing magical practices drawn from the victim's cultural background. In many cases, this type of control requires no systematic use of violence (Abbatecola, 2002: 93–95; Roversi, 2001: 35–54).

Other Reference Models

Other modalities of prostitution organizations, different from those discussed so far, can also be seen in the national context. Among these, Chinese prostitution seems to show certain specific aspects. What is more, this activity is performed indoors in locations such as apartments or so-called massage parlours. This type of sexual service may be connected to the cultural conditioning of Asian populations who tend to stigmatize street prostitution accessible to everyone (Light, 1977: 464–479).

The usual system for attracting customers is based on inserts in Italian periodicals. Messages take forms such as: "Chinese masseuses available. Nice-looking girls in their twenties give relaxing complete massages" (Tribunale Milano, 2002: 2). Other Chinese nationals manage customer appointments and prices range from 20 to 50 euros per session. Women who have chosen this line of work are paid between 5 and 10 euros; the rest goes to the criminal group. Otherwise, with women who arrive in Italy in debt to the trafficking organization, exploiters (male and female) take all the profits and leave the girls only the minimum necessary to survive (Abbatecola, 2002: 79).

Overall knowledge of the methods adopted by exploiters of Chinese women is limited. Even though a certain increase has been recorded over the past few years, from a statistical viewpoint, Chinese involvement in this crime is still low: only 56 people were brought to justice between 1995 and 2003.[6] This small

---

[5] The presence over the past few years should be noted of groups of Romanian exploiters who adopt enslavement strategies based on the systematic use of violence, the same as those used in Albanian criminal structures (Transcrime-Dna, 2004: 173); interview with a trafficked Romanian girl, Milan, 20 April 2003.

[6] At the end of 2003, 100,109 Chinese citizens were legally residing in Italy. Even taking into account undocumented Chinese immigrants, recorded involvement in the exploitation of prostitution still turns out to be very low (Caritas, 2004: 135).

presence in judicial statistics can partially be traced to the specific nature of Chinese crime that operates fairly exclusively inside their own community, thus creating notable difficulties in identifying illegal events. In fact, in many cases the penetration of this *criminal universe* was possible only thanks to the collaboration of Chinese citizens themselves. Many judicial investigations that have helped uncover kidnapping and extortion inside the Chinese community began subsequent to the victims' decision to turn to the Italian authorities (Becucci, 2001: 60–128). What is more, organizing prostitution in isolated places, hidden to the public eye, makes it more difficult for the police to identify the sort of exploitation concealed inside. Nevertheless, even considering these aspects, the limited Chinese involvement in sexual exploitation seems to be tied to other factors. The type of emigration based in great part on the family nucleus and the existence of informal social controls inside the Chinese community constitute the main block to the development of any substantial offer of commercial sex services. It is probably no coincidence that women who work as prostitutes are usually not from the two geographic-linguistic groups in Zhejiang and Fujian, the most numerous and organized among the Chinese immigrants in Italy, but from areas in the north of China such as Manchuria, whose inhabitants have undergone sudden impoverishment because of the privatization of state businesses and drastic reduction in social assistance (Abbatecola, 2002: 77). Frequently alone, lacking in mutual aid networks, these women turn to prostitution since the only other alternative is working for a few euros a day in a sweatshop run by fellow Chinese.[7]

Finally, the presence should be noted of groups of exploiters from countries in Latin America. In many cases, this involves prostitution in apartments. Usually, the woman contacts customers by distributing announcements in local newspapers and receives them at her house (Questura Lecce, 2000: 5). While Chinese prostitution tends to assume entrepreneurial characteristics (in some cases, women manage "massage parlours" alongside men), Latin American prostitution leans toward the typical model of the "kept" Italian pimp; the exploiter lives with the woman in the same apartment, furnished with a meeting place for clients, spends his time at the local bar and lives off the prostitute/lover.

---

[7] Recent research carried out in Prato, a city 10 km from Florence with around 20,000 Chinese inhabitants out of an overall population of 150,000, stated: "Chinese immigrants from Zhejiang maintain their women could never become involved in this activity [i.e., prostitution: Author's Note] because they have too many relatives and acquaintances in Italy and fear the judgment of the entire community (...) On the other hand, women from Dongbei, [former Manchuria: Author's Note] would not have this obstacle since they normally emigrate alone and are either unmarried or divorced" (Tolu, 2003:147).

Italian Mafia-Type Associations and Foreign Criminal Groups

Although judicial statistics show a noticeable presence of Italians in the exploitation of prostitution, no equally significant involvement of indigenous mafia organizations emerges from an analysis of trial records.[8] Neither do any important relationships emerge between mafia associations and foreign groups. Instead, Italians involved in the proceedings examined are almost always individuals connected to foreign exploiters in different ways. In other cases, they directly exploit the women within those "typically" ambiguous relationships pimps adopt with prostitutes.

In general, mafia-type associations appear uninterested in directly controlling this market. Aside from preferring more lucrative illegal activities, such as drug trafficking for example, motives for their lack of interest should be sought in the traditional, cultural reluctance in some mafia brotherhoods. For example, this is the case with Cosa Nostra affiliates who consider earning profit from sexual exploitation reprehensible. According to collaborator Antonino Calderone, "The mafia does not run prostitution because it is a dirty business ... Francesco Rinella, the brother of two *uomini d'onore* (men of honour) and son and grandson of two *uomini d'onore* was never admitted into Cosa Nostra precisely because word had it he was a *magnaccia* (pimp)" (Arlacchi, 1992: 5). The reluctance of Cosa Nostra to manage the prostitution market was recently confirmed by the Public Prosecutor's Office in Palermo that excluded any involvement of this syndicate's members in the sex trade (Transcrime-Dna, 2004: 173).

Though elements pertaining to this specific mafia culture have a certain importance, what could be said on the other hand about members of the Camorra who have never been bothered by moral scruples? In Naples, up until no more than 15 years ago, around 10,000 people were employed in the prostitution business; but over the past few years, no trace has been left of the Camorra's involvement in this activity (Direzione Distrettuale Antimafia, 2002; Lamberti, 1992: 32).

Those who control the traffic of women from their countries of origin, such as Nigerian and Chinese organizations, or those who, like Albanian traffickers, use recruiters in the area of the Balkans, have an immediate advantage not easily equalled by other actors, Italian or foreign, outside this criminal network.

Though not directly present in the management of the sex trade, mafia-type associations nevertheless exercise forms of indirect control over the exploiters. There is some confirmation of this: along the seafront between Naples and Caserta,

---

[8] Sixteen reports presented to judicial authorities were examined of women trafficked and forced into sexual exploitation and twenty-six judicial proceedings initiated by Public Prosecutors in the Italian cities of Brescia, Milan, Torino, Florence, Lecce, and Naples from 1998 to 2003. Because of the limited space available, only those judicial proceedings cited in the text are reported.

Camorra groups demand a sort of tax from prostitutes and their exploiters in order to set up shops along the roads (Dda, 2002: 37). In the same way, certain 'Ndrangheta *cosche* (gangs) have allowed foreign groups from Eastern Europe to run prostitution in exchange for drugs at a special price brought into Calabria by the foreign exploiters (Macrì, 2004: 5). Finally, some members of the 'Ndrangheta have exploited women working as prostitutes in their hotels in Milan. Nevertheless, in this case, the organization as such was not running the exploitation, but rather individual members who, either more or less tacitly, were not hindered by their clan bosses (Tribunale Milano, 2001: 146).

Conclusions

In summary, we could say a real criminal succession has occurred in the street sex market over the past 10 years. Italian pimps, as well as exploiters with ties to organized crime, have mostly abandoned this activity.[9] The inclusion of new illegal actors has moved in step with the arrival of foreign women who have progressively replaced Italians in offering sexual services for payment. The streets have become populated with foreign prostitutes whose exploiters are their own countrymen or other groups of foreigners.

The lack of interest by mafia organizations has no doubt facilitated the prevalence of new actors in this lucrative business. Nevertheless, the motives that have encouraged this turnover to foreign groups can be found in the fact that the migratory chains revolving around prostitution have escaped the control of Italian mafia-type organizations. Women destined for the Italian market arrive from Nigeria, Albania, and various countries in Eastern Europe. Those criminal groups capable of setting in motion the offer of sexual services on an international scale tend also to control, though with some major differences, the final exploitation market in Italy.

Over the next few years, we shall see whether or not the illegal proceeds amassed from the prostitution business will permit the best-organized foreign criminal groups to acquire positions in other more profitable illegal activities, such as drug trafficking, great enough to constitute a potential threat to the traditional criminal hegemony of indigenous mafia associations.

References

Abbatecola, E. (2002). Le reti insidiose. Organizzazione e percorsi della tratta tra coercizione e produzione del consenso. In M. Ambrosini (Ed.), *Comprate e vendute. Una ricerca su tratta e sfruttamento di donne straniere nel mercato della prostituzione* (pp. 69–133). Milano: Franco Angeli.

---

[9] Where they are present in managing the sex trade, Italian exploiters are prevalently involved in "hidden" market segments such as prostitution in apartments and "night clubs."

Arlacchi, P. (1992). *Gli uomini del disonore*. Milano: Mondadori.

Becucci, S. (2001). Il fenomeno criminale cinese in Italia: caratteristiche e aspetti problematici. In S. Becucci & M. Massari (Eds.), *Mafie nostre, mafie loro. Criminalità organizzata italiana e straniera nel Centro-Nor* (pp. 68–120). Torino: Comunità.

Bufo, M. (2005). Prostituzione, trafficking, sommerso: linee evolutive di un fenomeno. In AA. VV, *Il sommerso. Una ricerca sperimentale su prostituzione al chiuso, sfruttamento e trafficking* (pp. 10–16). Quaderni di Strada 2.

Carchedi, F. (2000). La prostituzione straniera in Italia: analisi dei risultati dell'indagine sulle protagoniste e i modelli relazionali. In F. Carchedi et al. (Eds.), *I colori della notte* (pp. 100–162). Milano: Franco Angeli.

Carchedi, F. (2003). Le modalità di sfruttamento coatto e la prostituzione mascherata. In F. Carchedi, G. Mottura, & E. Pugliese (Eds.) *Il lavoro servile e le nuove schiavitù* (pp. 125–146). Milano: Franco Angeli.

Caritas. (2004). *Immigrazione. Dossier statistico 2004*. Roma: Nuova Anterem.

Corso, L., & Landi, S. (1998). *Quanto vuoi? Clienti e prostitute si raccontano*. Firenze: Giunti.

Dal Lago, A., & Quadrelli, E. (2003). *La città e le ombre. Crimini, criminali, cittadini*. Milano: Feltrinelli.

Direzione Distrettuale Antimafia (Dda). (2002). *Relazione sullo stato della criminalità organizzata nel Distretto Giudiziario di Napoli*. Napoli: Procura della Repubblica.

Donadel, C., & Martini, R. (Eds.) (2005). *La prostituzione invisibile*. Regione Emilia Romagna: Progetto West.

Istituto Nazionale di Statistica (Istat) *Annuario Giudiziario (Anni: 1995–2003)*, Roma.

Lamberti, A. (1992). *La camorra. Evoluzione e struttura della criminalità organizzata in Campania*. Napoli: Boccia.

Light, I. (1977). The ethnic vice industry. 1840–1914, *American Sociological Review, 42, June*, 464–479.

Macrì, V. (Ed.) (2004). *Conferenza nazionale sulla 'ndrangheta*, Direzione Nazionale Antimafia, Roma, 12 May (pp. 1–44).

Massari, M. (2003). Il mercato della prostituzione straniera. In M. Barbagli (Ed.), *Rapporto sulla criminalità in Italia* (pp. 301–323). Il Mulino: Bologna.

Monzini, P. (2002). *Il mercato delle donne*. Roma: Donzelli.

Omicron. (2001). *L'influenza della criminalità straniera sulla struttura degli interessi e dei comportamenti criminali: le grandi aree metropolitane nell'Europa mediterranea (Barcellona, Parigi, Milano*. Osservatorio Milanese sulla criminalità organizzata al nord, Milano, Programma Falcone: 2000/Falcone/183.

Questura Lecce. (2000). Informativa di reato, 28 July 2000.

Questura Udine. (2000). *Annotazione finale sulla Mafia Nigeriana. Operazione Edo, 18 February*.

Roversi, A. (2001). La prostituzione da strada delle donne immigrate a Modena. *Polis, XV*(1), 35–54.

Taylor, I., & Jamieson, R. (1999). Sex trafficking and the mainstream of market culture. *Crime, Law and Social Change, 32*, 257–278.

Tolu, C. (2003). Diversificazione nei luoghi di origine dei migranti cinesi. In A. Ceccagno (a cura di), *Migranti a Prato. Il distretto tessile multietnico* (pp. 137–166). Milano: Franco Angeli.

Transcrime-Direzione Nazionale Antimafia (Dna). (2004). *Tratta di persone a scopo di sfruttamento e traffico di migranti*, www.giustizia.it. *Judicial Files*

Tribunale Genova. (2001). Ufficio del Giudice per le indagini preliminari, Ordinanza di custodia cautelare in carcere, 2 November.

Tribunale Lecce. (2000a). Verbale di incidente probatorio, 24 February.

Tribunale Lecce. (2000b). Ufficio del Pubblico Ministero, verbale di denuncia orale, Lecce, 8 March.

Tribunale Milano. (2001). Sentenza nei confronti di Abazi Enver +49, 13 April.

Tribunale Milano. (2002). Richiesta di rinvio a giudizio contro Zhang Yan +2, 15 April.

Chapter 6
The Vehicle Theft Market in Bulgaria

Tihomir Bezlov and Philip Gounev

In the United States, the FBI reports that in 2005, 1.2 million vehicles were stolen at a value of $7.6 billion (Federal Bureau of Investigation, 2005), while in Europe, Europol estimates that in 2004 approximately one million vehicles were stolen, of which at least 450,000, at a value of €6.75 billion, were never recovered (Europol, 2006). Despite the significant financial impact that vehicle theft has on victims, it has remained an area with very little academic or policy research. In some respects vehicle theft is just an aspect of property crime. Most victims, however, consider it a serious crime, and in the EU 92% of car-thefts are reported to the police (EUICS Database).

On the other hand though, car theft is often associated with serious organized crime activities and therefore has attracted special attention from law enforcement authorities across Europe. The complex level of criminal organization involved, and the use of corruption, fraud, and even violence are more reasons to pay close attention to the impact the car-theft market has on victims and law enforcement authorities, particularly in the new EU member states. Adding to this is Europol's forecast of the growing importance of car-theft as improved anti-theft devices will cause criminals to resort more often to violent robbery or burglary (Europol, 2006: 28).

Understanding the mechanics of stolen vehicle markets in countries such as Bulgaria is important for several reasons. First, Europol (2006: 7) highlights the growing role of Bulgarian-run vehicle theft crime groups across Europe. Second, knowledge of the stolen cars market in countries such as Bulgaria helps in drawing a full picture of the stolen vehicles market in Europe, by including what is traditionally considered a destination country. One should keep in mind that the patterns of the car-theft market in Bulgaria are quite different from most EU member states. In Bulgaria ransom seeking, corruption, the communist legacy, and the role of mafia racketeering that gripped economic and political life in the 1990s, all make for a unique mix of underlying factors that shape the present-day stolen car market.

Thirdly, the description of the mechanics of the stolen vehicle market in a new member state would allow policy makers to design better common EU instruments for combating transborder car theft in the EU.

Lastly, the present article tries to bridge a gap in the existing literature on car theft in Europe. Clarke and Brown's (2003) overview of the knowledge gaps about international trafficking of stolen vehicles well demonstrates the need for further research in this area. The only other study that the authors came across was conducted by

D. Siegel and H. Nelen (eds.), *Organized Crime: Culture, Markets and Policies.*
© Springer 2008

Kilias and Gerber (2003), who analyzed the stolen vehicle market in Europe and Russia (Kaliningrad). Other studies (Europol, 2006) generally do not address the motor vehicle theft markets in Eastern Europe, but focus on the specifics of the countries of origin of stolen vehicles. The numerous Home Office studies on car theft in the UK, such as Brown (1995), Brown and Billing (1996), Brown and Saliba (1998), and Houghton (1992), focus on specific aspects of UK policies and do not address the East European aspects of this type of crime. In addition, their methodologies do not go beyond examination of law-enforcement databases or victimization surveys, and do not address organized crime aspects in any considerable depth.

This paper will first present the somewhat unique background of the present-day car-theft market in Bulgaria by reviewing its structural development during the 1990s. It will then focus on analyzing the networks that characterize two aspects of the car-theft market – the theft for ransom and the market for trafficked vehicles from Europe. While the first aspect primarily affects the average Bulgarian, the second aspect concerns the majority of EU citizens. We have focused most of our research on the theft-for-ransom market, but we have also examined some issues related to the trafficking of stolen vehicles – such as the role of Bulgarian crime groups in the EU and the mechanics of trafficking vehicles into Bulgaria.

The findings in the paper are the results of a larger study on organized crime trends in Bulgaria, carried out by the Center for the Study of Democracy.[1] For the purposes of the present research, close to fifty interviews in five towns were carried out during 2006 and 2007, in addition to analysis of police statistical data and victimization surveys. The interviews included meetings with law enforcement officers (police, border police, customs, and organized crime units), insurance company representatives, car thieves and other individuals involved in the auto theft market. For the sake of the interviewees' security, we do not mention names, places or interview dates. We use the nicknames of some of the actual criminals that are either publicly known from media reports or were revealed by the interviewees.[2]

Historical Background

The Early 1990s

The history of the vehicle theft market in Bulgaria to a large extent reflects the history of organized crime in the country. Until 1990, the purchase of private cars in Bulgaria was extremely restricted. The average wait time to purchase a car

---

[1] The full report can be found at http://www.csd.bg.

[2] We adhere to nicknames instead of legal names for liability reasons, as very few of these individuals have been convicted in court.

was 6–10 years. In addition, it required 10–15 years of savings from an average salary, while no financing schemes existed. Therefore, there were strong pull factors for a market for stolen cars. Nevertheless, the strong police control and the close to 100% recovery rate of stolen vehicles left only a handful of deviant youth to be involved in joy-ride thefts. Between 1980 and 1985, the annual car thefts were only about 2,000; after 1985 the picture started to change (see Fig. 6.1).

In 1990, many car thieves were among the 4,000 prisoners that were amnestied. Some of them headed to Western Europe to become involved in a range of criminal activities, including auto-theft, while others settled in the Czech Republic or Hungary, where they became intermediaries or itinerant groups that facilitated trafficking of stolen cars. These individuals' technical skills and connections with the Balkans and the Middle East provided the key competitive advantage that allowed them to compete with local and former Soviet Union crime groups. By 1992, a large group of car thieves, including the late Ivo Karamanski, later to be known as the "godfather of Bulgarian mafia" (as well as a few others that today are "prominent" car thieves), were repatriated to Bulgaria by Hungarian and Czech authorities. This accelerated even further the growth rate of stolen cars, while at the same time these individuals had developed the connections to import stolen vehicles from Western Europe.

The Bulgarian law-enforcement authorities were quite inexperienced and lacked the operational capacity to deal with the complex networks involved in the trafficking of stolen vehicles from Western Europe. In the early 1990s, the car became a status symbol in Bulgaria, thereby creating a considerable demand for cheap but luxury

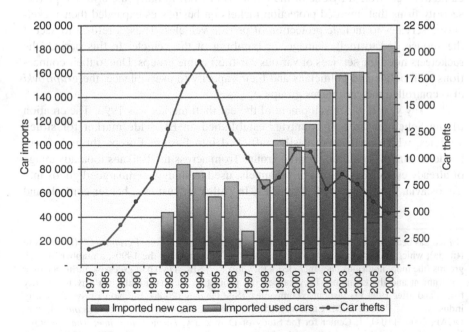

Fig. 6.1 Stolen cars and the car market. *Source*: Ministry of Interior

cars, as loan or leasing schemes were nonexistent. This demand was met primarily by stolen luxury vehicles from Western Europe.

During that period, there were two markets for stolen vehicles. One was for car parts – as the decline of state economic ties with the former communist countries created a deficit for spare parts, until private trade relations filled this vacuum. The second was the resale market – which became possible due to institutional crisis and corruption that permeated all law enforcement institutions in Bulgaria. The import and domestic sale of stolen vehicles turned into a widespread practice as corruption allowed unimpeded import and registration of stolen vehicles. Networks comprised of car-thieves, drivers, used-car dealers, customs, police, and border police officers dominated the stolen vehicle market in this early period. In the year 1992, 12,711 vehicles were registered as stolen, but according to police officers, the real number was at least twice as high.

Enter the "Grupirovki"

It was only after 1992 that a third market appeared, similar to what Killias and Gerber (2003: 219) observed in Russia and called the "theft for ransom" market. This coincided with the takeover (1992–1994) of the stolen and trafficked vehicles market by the major mafia-type crime groups, known as "grupirovki" in Bulgaria.[3] During this period the market for stolen luxury cars, as well as most car thieves, came to be controlled by one of the main grupirovki. Initially, the criminal private security firms that imposed protection rackets on businesses expanded their racketeering services to include protection of private vehicles. Thus, a refusal to accept the service, led to theft, damage, or bombing of the vehicle. In this period, the racketeers hired the services of various car-theft crime groups. Due to their connections with police and politicians and their capacity to use violence, the grupirovki also controlled the car-thieves groups.

The key year in the development of the car theft market was 1993. The creation of the Schengen area of countries established an EU-wide market for stolen vehicles, which were more easily trafficked into Eastern Europe through fewer border controls. In addition, crime groups from across the Balkans took advantage of already established trafficking channels, used to smuggle embargoed goods into Yugoslavia, to smuggle stolen vehicles. In Bulgaria, customs, border-controls and

---

[3] Much has been written on mafia-type groups in Bulgaria, called "silovi grupirovki" (identical to Russia), which roughly translates as "power groups." For most of the 1990s, a number of such groups (the more notorious ones included VIS, Club 777, SIK, TIM) had excessive influence over government and the economy. They carried out protection rackets initially via private security firms, and after 1994 via insurance companies. (See Gounev (2006). Bulgaria's private security industry. In A. Bryden & M. Caparini (Eds.), *Private actors and security governance*; LIT & DCAF, 2006: 110–114; Center for the Study of Democracy, *Partners in crime: The symbiosis between the security sector and organized crime in Southeast Europe, Sofia,* CSD 2004).

police were largely controlled by the *grupirovki* (many of whose members were former law-enforcement officers).

Domestically, two developments influenced the size of the market. First, the *grupirovki* gradually spread their power around the country, which enabled them to guarantee security of insured vehicles nationwide. At that time, the VIS-1 *grupirovka* acquired national "coverage." It confronted or allied with smaller regional players, such as Group 777 in southern Bulgaria, TIM in the port city of Varna, and First Private Militia in the port city of Bourgas.

Secondly, by 1994 the *grupirovki* were forced to restrain racketeering related violence, and switched to a tactic where car-theft became an important instrument to extort businesses and individuals. At that point the theft of vehicles for car-parts constituted only 20% of the total cars stolen, while ransom seeking and racketeering became the most widespread reason for car-theft. Victims received a phone call within 20 min after the car was stolen, even before they had time to contact the police.

Ransom-seeking and racketeering took on such vast proportions that the share of stolen cars went up to 20–25% of all imported vehicles (according to official data; in reality it was over 30%). The regular insurance companies were not willing to provide car insurance, due to lack of security, which opened an opportunity to transform the private security firms into insurance firms, where insurance became tantamount to a new form of the old protection racket. In 1994, VIS-1 was forced by the government to close down, but it re-registered as an insurance company called VIS-2. Because VIS-2 controlled many car-thieves, their insurance coverage usually amounted to ordering the thieves to return the stolen car, and on rare occasions paying for it. Therefore, the VIS-2 insurance sticker signalled to the thieves that this car was protected by VIS-2.

In early 1995, some of the VIS-2 members departed to form an alternative insurance racketeering *grupirovka*, called SIC (whose former members even today exert undue influence in business and politics). Within a year, the car-insurance and the related car-theft markets were split between the two companies. Their stickers, indicating that they had insured the property, became ubiquitous not only on cars, but also on offices, stores, street kiosks, and even public transport stops. The two companies hired a large number of insurance professionals, who worked side-by-side with teams of hit-men that specialized in punishing disobedient car-thieves, or "finding" the car (and delivering a new one, often a trafficked car, in case the car that was originally stolen could not be found). In 1996, smaller regional splinter racket-insurers appeared (such as Apolo and Balkan, Korona Ins., Levski-Spartak, Zora Ins., etc.). Certainly not all players (thieves) were brought under the control of the above companies.

Between 1995 and 1996, the import of stolen cars from Western and Central Europe continued to grow. Dozens of crime groups in Western Europe were trafficking stolen cars to Bulgaria (particularly the higher class models that were in great demand), using the range of approaches described in Europol (2006). After the stolen cars were sold to Bulgarian customers and insured by the *grupirovka*, they were often stolen again and resold or exported to the Middle East or elsewhere in the Balkans.

All of these schemes worked as a result of the pervasive corruption within the national and border police forces, facilitated by the law-enforcement background of key *grupirovki* members. Other factors included the continued control over local car-thief groups and access to the political and economic elites who acted as both buyers of high-class stolen vehicles and protectors of the criminals involved.

The new order on the market imposed and controlled by the VIS-2 and SIK insurance companies gave quick results, as in 1995 the number of stolen cars fell by 12% and by another 26% in 1996. Based on the 1995 figures for stolen cars and data from the insurance market in that year, the income generated in the stolen car market was around €80 million, which at the time was about 1% of the country's GDP.

The beginning of the end of the racket insurance companies came with the political changes and the new government in 1997, which marked the end of the economic and political crisis and the hyper-inflation that had gripped the country during 1996. By 1998, all insurance companies had to be re-registered and the specific provisions in the new Law on Insurance banning private security services by insurance companies, as well as a new political climate, led to the gradual abandonment of the racket insurance practices.

In that year, the uncertainties surrounding the future of the *grupirovki*, and the anti-corruption measures in law-enforcement institutions brought about a significant drop in the number of stolen cars to the 1991 levels. All racket insurance companies named above were under pressure by the government and about to lose their insurance licences. Their infamous stickers (that indicated the insurer of the property) were banned, and tax auditors and law-enforcement officials raided their offices repeatedly. The new government dismantled the existing networks by replacing all high-level police, tax, and customs officials, while many lower-level thugs were detained.[4]

During that period not a single *grupirovka* leader was prosecuted, as the politicians chose to reach an agreement with them. In the following 2 years (1999–2000), although VIS left the insurance market, SIC managed to transform some of its business, and as a result car-thefts started to increase again. The increase, though, was also partly due to the fact that the newly established insurance companies did not make pay-outs unless the car-theft was registered with the police, which led to an increase of police-registered car-thefts. Adding to this was a 1999 prison amnesty, which saw about 2,000 criminals, including car thieves being freed (Bezlov, Gounev, & Hristov, 2006: 29).

The Post-2001 Market

The present-day car-theft market is best understood if one takes into account some of the developments that took place between 2001 and 2002. First, after 2001 there was a rapid increase in the number of imported cars, as they almost doubled

---

[4]Between 1998 and 1999, the number of detentions in the country almost doubled from 6,478 to 10,785, according to data from the Bulgarian Ministry of the Interior.

between 2001 and 2005, from 100,000 to 180,000. This over-supply brought the prices to such low levels that the incentive to purchase stolen vehicles decreased significantly, except for the highest-class cars. Secondly, in the period 2002–2003 border and customs controls were modernized through the introduction of institutional systems connected to national and international databases that allowed for quick identification of vehicles that were reported stolen in Western Europe or in Bulgaria. The number of thefts in Western Europe reached a 10 year-low, falling from 2.07 million in 1993, to 918,000 in 2004 (Europol, 2006).

The ransom-seeking activities gradually started to decrease, as insurance companies started paying out for stolen vehicles within 3 months (down from 6 to 9 months), and the victims were less inclined to pay ransom. Improved anti-theft and tracking devices also became widespread. Police reform also played a role, as the staff of vehicle crime teams across the country was increased, which in turn reduced its vulnerability to corruption by car thieves. Also, administrative measures made the return of stolen vehicles much quicker, which eliminated the practices of paying bribes to police or prosecutors to return a vehicle which used to be kept as evidence while a trial against the thieves dragged on. Finally, the removal of EU visa restrictions for Bulgarians in 2001 led to the emigration of a significant part of Bulgarian criminals, including car thieves (Bezlov et al., 2006: 32–34). All of the above brought about a sustained reduction in car-theft: an average reduction of 12.8% for the 5 years to 2006, according to police data.

The above trends led to significant transformations: the theft and trafficking of stolen vehicles became riskier and more difficult, and ransom-seeking developed into a complex criminal network resistant to disruption.

The Present-Day Vehicle Theft Market

Our interviews with law-enforcement officers and participants in the criminal networks revealed that the old hierarchical mafia-like model has disappeared. The protection by *grupirovki* of car-thieves no longer exists. The new market is much more segmented and localized. While 60% of car thefts take place in the capital Sofia, the rest are largely concentrated around the three other large regional centres of Plovdiv, Varna and Burgas.

To illustrate how the market functions, we will describe in greater detail the network structure of the car-theft market in the capital Sofia, focusing on the ransom-seeking scheme. The individuals listed in Fig. 6.2 were the actual participants (for legal purposes referred to only by their nicknames) at the beginning of 2007.

The market has a strict division of labour. The first important set of players consists of the *car thieves groups* (such as the ones headed by Ivan the Fat or the Duck). These groups often have a narrow specialization, such as stealing only luxury vehicles, or mid-class cars (VW, Opel, etc), or trucks and vans, or Russian-made cars. They use equipment ranging from simple slim-jack instruments to electric-shock batons and more sophisticated devices. Coded keys are overcome by decoding

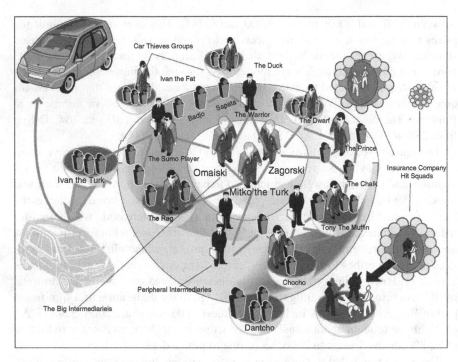

Fig. 6.2 Organization of the ransom-seeking market in Sofia

them; GPS tracking systems are removed or temporarily disabled via electromagnetic devices; entire electric and computer panels are changed, etc. Sometimes several vehicles participate in an operation to make sure that the escape routes are clear. On some occasions, a Jeep class truck follows the stolen car in order to push or tow it in case the anti-theft devices block the car engine. The car is then moved a few kilometres away to a parking lot, or a rented garage (for more expensive vehicles), where it stays until the ransom is received or the car is resold. Incidents are rare, as the police rarely risks engaging in movie-like car chases.

The other major set of players is formed by the range of *intermediaries* who are involved in requesting the ransom from the victim and arranging the return of the stolen vehicle. The small world principle (Milgram, 1967: 60–67) is successfully used and on most occasions the thieves are put in indirect contact with the victim. This process is aided by other players such as car-mechanics, police officers, used-car dealers, etc., as well as by the cultural importance that connections carry in Bulgarian society.

After stealing a vehicle, the car-thieves usually contact one of the three (known) *big intermediaries* in Sofia (Mitko the Turk, Zagorski, or Omaiski). They neither see the cars nor the ransom money but are central to the network operations, while very difficult to investigate, as they use multiple prepaid phone cars and operate largely while driving around in a vehicle. Trust is central to their role and to the overall functioning of the market. They guarantee that the victims receive their vehicle while the thieves get their payment. Violence is rarely used. Due to the relatively small number of thieves and intermediaries one risks being pushed out of the market if trust is betrayed.

The victim is rarely contacted directly, but rather through relatives, friends, or acquaintances who are approached by the intermediaries. Often, it is the victims who seek out contact by using their network of acquaintances. The ransom and the car-return also often take place via these individuals, preferably someone whose car has been stolen before. The victim's contact person is usually not directly approached by the intermediary, but by someone he/she might know – a car mechanic, a local cop, or a locally known crime-figure (who gets a commission of between €50 and €100).

If the negotiations for the ransom are unsuccessful, a couple of outcomes are possible. Cheaper cars are abandoned, while more expensive ones are sold for parts or trafficked out of the country.

In other words, the payment of the ransom passes through 3–4 layers of individuals before reaching the thieves, who are left with an average of €500 in profit, depending on the value of the stolen car. Investigation is difficult, and usually the first-line individual will claim to be assisting the victim, while other individuals in the money-transferring chain are difficult to trace. One tactic used is to quickly exchange the currency of the money in designated exchange bureaus, since the police often use marked bank notes to gather evidence. Another difficulty in investigating such cases is the heavy use of coded language and specific slang by thieves and intermediaries, which would not be accepted as evidence in a court of law.

In addition to the players described above, a separate role is still played by *insurance companies*. According to the interviewees as well as industry sources, there are presently three insurance companies that have unofficial units which use violence and informal negotiations to return stolen cars.[5] These units are not formally part of the company, and could be registered as separate firms but they solely work for one company and exclusively focus on cars. These three companies control about 50% of the car insurance market. They are connected to individuals or companies that were involved with racketeering in the past. Much of these companies' business is outside Sofia, where finding the actual thieves and the vehicle is much easier. In Sofia, the insurance companies face the same problems as the police: the multiple intermediaries conceal the perpetrators.

Therefore, the capital remains somewhat of a safe haven for car-thieves, with 65% of the vehicles being stolen in Sofia, while less than 50% of all vehicles in the country are registered there. In addition, car-thieves often prefer to move to Sofia because in small towns the police and insurance companies have influence over the judiciary and members of local criminal groups are more likely to face conviction.

The scale and the financial impact of the ransom-seeking vehicle thefts is clearly revealed in the National Crime Survey (NCS), carried out annually by the Center

---

[5] Due to possible liability claims they are not named here.

for the Study of Democracy. The NCS showed that 30% of the victims of car theft in 2005 were asked for a ransom, 56% of whom paid the requested amount.[6] The interviewed police officers felt that the majority of the victims were asked for a ransom (as high as 60%) and that 90% of them eventually paid up. They argued that many victims feel threatened and do not admit to having been asked for a ransom in surveys. The NCS indicated that in 2004 and 2005 the average cost of a stolen vehicle was around €3100. The average ransom asked, according to the victims, was €1100 (Bezlov et al., 2006: 56). One should keep in mind that the average monthly salary in Bulgaria is around €250 per month. Depending on whether one uses the NCS or the police estimates in 2006, the total amount of ransom money paid was between €500,000 and €3.8 million.

Trafficked Vehicles

The interviews with car-thieves indicated that the trafficked vehicles market is still significant, although its real size is difficult to estimate. Unlike victimization surveys and registered car thefts, there are few measures that provide an indication as to the actual number of trafficked vehicles. For some types of luxury vehicles, such as the BMW X5 sport utility vehicles, the interviewees (both police officers and car thieves) claimed that the number was over 30%. Such claims, however, are difficult to verify.

The smuggling and registration of stolen cars is now much more difficult than in the mid 1990s. The main tactic to circumvent law-enforcement measures is through the use of "cloning." The pull factors are usually used-car dealerships and used-car importers who are approached by clients willing to take the risk of buying a stolen car, for a price that is 10–20% lower than normal. To avoid attracting attention, the quantities ordered by dealerships are usually small. The second phase is the ordering of the vehicle from a group of car thieves operating abroad, usually in Spain, Italy, or France.

The third phase entails finding the vehicle. The groups make sure that the stolen vehicle is perfectly legal. According to Europol and public statements by local police forces in West European countries (Europol, 2006), robberies and burglaries with the aim to steal keys is usually the preferred way to overcome the complex security mechanisms of modern luxury vehicles. Then, a second vehicle (preferably of the same brand and model), usually from a junk yard or a used-car dealership is found, and all registration numbers (on the engine, chassis, etc.) are taken. Once the vehicle is stolen, the numbers from the used/junk car are copied onto the stolen vehicle, which then becomes a clone. The cloning of the numbers is usually done so meticulously that it can only be discovered with special instruments.

---

[6] The NCS uses the basic structure of the International Crime Victims Survey (ICVS) methodology and questionnaire.

Conclusion

The above analysis shows that the stolen vehicle market in Bulgaria has changed and has become smaller as a result of a multitude of domestic and international factors. Nevertheless, the clear-up rate for stolen vehicles in Bulgaria remains around 15%, while the average for the EU is around 40%. As Kilias and Gerber suggest, this indicates a strong presence of organized crime.[7] The scale of this presence is also well substantiated by crime victims surveys such as the NCS, which indicates that at least 30% of the victims of car theft in 2005 were asked for a ransom, 56% of whom complied with the demands. In view of the relatively low police capacity, the complexity of the networks, the difficulty in proving cloning, etc., there are several possible scenarios as to the future developments of the car-theft market in Bulgaria.

One possible scenario would be the continuing trend in the reduction of ransom demands, as more and more vehicles are fully insured. This scenario would be aided by the increasing electronic surveillance in Sofia and the integration of police and border databases with international databases. The increasingly sophisticated security measures by manufacturers and the growing number of official dealer-run car maintenance repair shops that have the capacity to work with sophisticated electronics, could reduce the number of grey-area repair shops presently involved in theft schemes. A more realistic scenario, however, would assume a continuation of the current situation, including a possible increase after the removal of border controls and the accession of Bulgaria to the Schengen Area. After all, Bulgaria being bordered by Turkey and the Western Balkans always leaves opportunities for trafficking of stolen vehicles. Finally, a more pessimistic scenario would assume that Bulgaria is simply experiencing a cyclical downturn of vehicle theft and is poised for an upward trend in the near future.

References

Bezlov, T., Gounev, P., & Hristov, H. (2006). *Crime trends in Bulgaria 2000–2005*. Sofia: Center for the Study of Democracy.

Brown, R. (1995). *The nature and extent of heavy goods vehicle theft*. Home Office Crime Detection and Prevention Series Paper 66. London: Home Office.

Brown, R., & Billing, N. (1996). *Tackling car crime: An evaluation of sold secure*. Crime Detection and Prevention Series Paper 71. London: Home Office.

---

[7] The recovery rate for stolen vehicles is a different matter. Since most vehicles are returned after the ransom is paid, the owners announce to the police that the vehicle has been "found" and the police registers the vehicle as recovered. Since the perpetrators are not found, the case remains not cleared. The Europol data does not make clear if European countries use the same standards to register recovered vehicles. Therefore the difference in recovery rates between countries seems unlikely. For instance, in 2004, Sweden's was 8.7% (the lowest in the EU), while Spain's was 69.5%. (Europol, 2006).

Brown, R., & Saliba, J. (1998). *The nature and extent of light commercial vehicle theft.* Crime Detection and Prevention Series Paper 88. London: Home Office.

Clarke, R. V., & Brown, R. (2003). International trafficking in stolen vehicles. In: M. Tonry (Ed.), *Crime and justice. A review of research volume 30* (pp. 197–227). Chicago: The University of Chicago Press.

Europol. (2006). *An overview of motor vehicle theft from a European perspective*, January 2006.

EUICS Database (http://www.europeansafetyobservatory.eu/euics_da.htm).

Federal Bureau of Investigation. (2005). *Uniform crime report 2005.* Washington: FBI.

Houghton, G. (1992). *Car theft in England and Wales: The home office car theft index.* Crime Prevention Unit Series: Paper 33. London: Home Office.

Kilias, M., & Gerber, J. (2003). The transnationalization of historically local crime: Auto theft in Western Europe and Russia markets. *European Journal of Crime, Criminal Law and Criminal Justice, 11*(2), 215–226.

Milgram, S. (1967). The small world problem. *Psychology Today, 2*, 60–67.

The Intertwinement of Illegitimate and Legitimate Activities

Chapter 7
Diamonds and Organized Crime: The Case of Antwerp

Dina Siegel

Though globalization is a term which is viewed increasingly more often in anthropological research as too abstract and vague, it still prevails in critical literature. In criminology, organized crime is presented as a number one example of the globalization process (in, amongst others, Fijnaut & Paoli, 2004; Galeotti, 2004; Mittelman, 2000). Globalization is usually defined in literature as a process which includes an extension of the market, a wide circulation of money, developed communication technology and world culture (Bauman, 1998; Beck, 2000; Eriksen, 2003). What criminology has brought to globalization studies is the recognition that crime in general and organized crime in particular is an inevitable part of the process. Galeotti distinguishes five main drivers for the globalization of organized crime: technological drivers ('from increasing the ease of travel to introducing new illicit commodities to be sold...', 2004: 4); political drivers, when organized crime is expected to respond to various political developments, for example wars, natural disasters or ethnic tensions (ibid:5); economic drivers, in which organized crime reacts similarly to its economic environment ('markets open and close', ibid:6); enforcement drivers, for example when operations of law enforcement agents are on the offensive and cause the perpetrators of organized crime to be more cautious or change their strategies; and internal drivers, namely changes in structures, for example, from 'monolithic' groups to 'loose networks of semi-autonomous criminal entrepreneurs' (ibid).

Usually organized crime operates in illegal markets which are regulated by the extortion activities of criminal groups. The most important condition for the existence of organized crime is its ability to keep its activities hidden. There are no written contracts or agreements in order to prevent the production of evidence of illicit transactions and it has its own internal system for solving conflicts, since legal institutions are inaccessible for solving disputes in illegal circuits. This internal system of rules, regulations, codes of behaviour and arbitrage makes it easy for organized crime to run a parallel economy, either in supplying illicit products or services (drugs, kidnappings, illegal sex-workers); taking advantage of local economic situations such as large price differences or product deficit (cigarettes, cars), or by illegally supplying licit products (weapons, exotic animals and plants, diamonds).

Despite long and continuous debates on what organized crime is, it remains a vague concept. What is considered organized (Chinese Triad or Japanese Yakuza) and what is not or disorganized (a group of teenagers pick pocketing in big cities);

D. Siegel and H. Nelen (eds.), *Organized Crime: Culture, Markets and Policies.*
© Springer 2008

how many members should the group include to be considered to be 'organized' (is a solo criminal who conducts bank fraud organized or not?); why are some specific activities viewed as organized (drug smuggling) and others not (cyber crime), which of these activities are illicit and which are not? In specific contexts, activities which are regarded as illegal in one country are not considered to be so in others (for instance, corruption). Criminologists have to deal with organized crime as an unclear and ambiguous concept. According to Paoli (2002: 52), the concept inconsistently includes two aspects: the supply of illicit goods and services on the one hand and a large-scale organization with a collective identity and inner division of work on the other. Paoli argues, however, that the supply of illicit goods and services is 'disorganized' rather than organized and that the famous large-scale organizations (such as Cosa Nostra, 'Ndrangheta, the Japanese Yakuza, the Chinese Triads, etc.) are not exclusively involved in activities on illegal markets (ibid). It is therefore important to keep these two aspects separate.

To make the matter even more complicated, another concept was introduced in the 1980s, namely 'transnational organized crime', which also rapidly became one of the favourite subjects in criminological research at that time. In their attempt to justify the emergence of the term 'transnational' in contrast with, for example, 'international' crime, some scholars suggested distinguishing between crimes which are 'acts prohibited by international criminal law on the basis of the 1994 draft code, multilateral treaties or customary practice by all nations' (Passas, 2002: 13), or in other words – international crimes, and in contrast, transnational (or cross-border) crimes which are acts which violate the laws of more than one country (Bossard, 1990).

Diamond-related crimes may be either local or transnational, varying from spectacular robberies from the big museums and exhibitions to smuggling through complex schemes from Sierra Leone to Antwerp. There are various criminal organizations involved in illicit diamond activities. In this article I will examine some aspects and forms of organized criminal activities. I will argue that if De Beers and various NGOs, which actively participated in the so-called 'anti-conflict diamonds' campaign, have contributed to raising the value of a diamond, the same is relevant for organized crime in promoting and continuing, often without even knowing it, the myth and glory associated with diamonds, which in its turn has a great influence on legal and illegal economic markets. In this context I will first analyse which criminal groups are involved in diamond-related crimes, second, in which illegal markets organized crime operates, and third, whether there is a link between illegal and legal activities in the diamond world.

The main questions which I will consider in this paper is whether the supply of the illicit 'conflict diamonds' in Antwerp is well-organized and whether large-scale mafia-type criminal groups are conducting criminal activities, such as spectacular robberies of diamonds in the city. In one way this could be considered to be an empirical test of the doubts expressed by Paoli regarding organized crime activities as mentioned above, based on the specific case of diamond-related crimes in Antwerp. I will focus here on the smuggling of 'conflict' or 'blood' diamonds from some African states on the one hand, and on large-scale robberies of the Antwerp Diamond Centre and Belgian airports, on the other.

Since the beginning of the 1990s, the diamond sector in Antwerp has been confronted with the rising interest of organized crime groups in diamonds in the context of smuggling, money laundering, organized robberies and fraud activities. The risk that organized crime will be able to penetrate the semi-autonomous sector of diamond traders, or the possibility that insiders will 'receive an offer they cannot refuse' and therefore become partners in crime will be illustrated here.

As maintained in official reports, conflict diamonds play an important role in the financing of civil wars and in the abuse and exploitation of innocent people. Various rebel groups smuggle these diamonds to Antwerp, after which they are traded throughout Europe and the US (Belgische Senaat, 2001). How do these 'blood diamonds' end up in the legal circuit? Which role do the 'certificates of origin' play, which are officially intended to keep the blood diamonds out of the legal sector? What is the involvement of organized crime in smuggling in Antwerp? I will also be considering how it is possible that organized crime succeeds in penetrating perhaps the most secure place in Antwerp, namely the Diamond Quarter. Is it possible for organized crime to operate without assistance from inside the diamond sector?

In this, I am basing my arguments on various documents from parliamentary committees from the Belgian Senate, on critical literature, NGO reports and on my own observations and interviews during fieldwork in Antwerp in the period from 2002 to 2005. This fieldwork included a wide range of research methods, from frequent conversations with the diamond traders, brokers and representatives of different diamond organizations and NGOs, to content analysis of international media, Internet reports and critical studies of different aspects of the diamond trade and crime in this context.

Smuggling 'Conflict Diamonds'

Smuggling diamonds is as old as the hills and it occurs in many places, often in periods of war and hunger and in countries with a troubled economy and instable government. But in times of peace smuggled diamonds are also a good investment and for some people, a guarantee of a better future. In contrast with drug trafficking or trading in rare species, trading in diamonds is legal. Naturally, smuggled diamonds are hidden from the view of customs and the police and when traders wish to evade taxes at the borders, smuggling can also be said to be taking place. But the smuggling which has attracted the most attention in recent years is the smuggling of conflict diamonds.

The concept of 'conflict diamonds' or 'blood diamonds' refers to diamonds which have been obtained from areas which are under the control of rebel groups in Sierra Leone, Angola, Liberia and the Democratic Republic of Congo. This concerns a very small percentage of all mined rough diamonds, namely 4–5% of the total world production (Tamm, 2002: 6–7).[1] In numerous NGOs reports, it was argued

---

[1] The 2005–2006 estimation was that, as a result of the Kimberley Process, conflict diamonds were reduced to 1% of world production.

that these diamonds allegedly play an important role in financing civil wars, and thus contribute to the suffering of the civil population.

In 1998, the British human rights organization Global Witness published a report in which the De Beers Firm was accused of purchasing diamonds from the Angolan rebel organization UNITA (National Union for the Total Independence of Angola) (Global Witness, 1998). The report was followed by the 'Fatal Transactions Campaign', a cooperative alliance by a number of human rights organizations (including Global Witness, Medio International, the Netherlands Institute for South Africa and Oxfam NOVIB). Two years later the Partnership Africa Canada brought out a report about the conflict in Sierra Leone, which described the way in which the rebels from the Revolutionary United Front (RUF) exchanged diamonds for drugs and weapons via neighbouring Liberia (Smile, Gberie, & Hazleton, 2000).

Since 1998, the UN has adopted a number of resolutions to bring a halt to the illegal trading of diamonds in the African countries in question. In 2001, representatives of the most important diamond producing countries and NGOs met in the South African city of Kimberley to develop a system to regulate the import and export of rough diamonds (the Kimberley Process). One of the most important decisions was that as of 1 January 2003, all the diamonds which are exported to the EU, have to be provided with a certificate of origin. The import and export of rough diamonds is only permitted when they are accompanied by a valid Kimberley Process Certificate (KP certificate).These diamonds have to be packaged in a fraud-proof container, as introduced by the authorities in the export countries. Moreover, trading in rough diamonds is only permitted between countries who participate in the Kimberley Process Certification Scheme (KPCS) for rough diamonds.

Fifty-two countries agreed to the KPCS. The support that the NGOs received from the various diamond organizations was striking, such as the *Hoge Raad voor Diamant* (HRD, the Diamond High Council), an umbrella organization of Antwerp diamond dealers, and in particular De Beers, the largest diamond multinational, who, with more than 70% of the world production of rough diamonds, has a near monopoly (*National Geographic*, 2002). There has been considerable speculation about the reasons for this support, but it is clear that if the diamond dealers had not participated in the campaign against blood diamonds, this would have led to a reduction in consumer confidence and thus constituted an economic setback for the sector. Up until now it would appear to be a success story: both NGOs and governments as well as the industry itself are working together in unison to exclude diamonds from areas of conflict in Africa.

However, since the start of the Kimberley process, various evaluation reports have appeared. There is doubt expressed in most of them about the effectiveness of the certificates. The American General Accounting Office (GAO) was the first to draw the conclusion that the certificate system did not work due to the closed and nontransparent nature of the diamond industry and because it is difficult to establish the origins of diamonds (GAO, 2002). Criticism has also been expressed in other reports and publications (e.g. see Peleman, 2002). One of the most important areas of criticism is the lack of reliable officials to monitor the origins of the diamonds and issue certificates of origin. The corrupt practices of civil servants are regarded by the local population as inevitable relations of patronage and clientism, which have their own morality (Shore & Haller, 2005: 12) and are rooted in a completely developed normative system (Ruud, 2000: 271). It is partially for this reason that corruption is difficult to combat in these countries.

But according to the experts the most important problem – and the Kimberley agreements cannot solve this – is smuggling (Belgische Senaat, 2002). The Kimberley process did not examine the mechanisms of the illegal trade in diamonds in depth. According to Peleman, an expert who gave evidence to the Belgian Senate, unregistered, smuggled diamonds 'are by definition not covered by the Kimberley agreements' (ibid). Rough diamonds are smuggled out of the conflict countries without that many problems and bought by dealers who sell them on to international trading centres in London, Antwerp or Tel-Aviv. Under the existing import control system it is extremely difficult to determine the actual origins of the diamonds.

In general, three aspects can be distinguished which enable smuggling from conflicts areas. Firstly, it is still impossible to determine the origins of rough diamonds on the basis of a mineralogical analysis. Secondly, false certificates are used, a new development in the activities of organized crime. Through this, it is possible to give a different location so that the origin of the diamond is provided with 'legal' status. Thirdly, diamonds are smuggled to neighbouring countries where there are no conflicts and illegal diamonds thus become legal, a form of whitewashing. According to Peleman, the certificate of origin threatens 'to become the ultimate whitewash system for conflict diamonds' (Belgische Senaat, 2001). Diamonds are smuggled through overseas territories, such as French Guyana, to Europe (Weyzig, 2004: 62). One of the strategies for whitewashing conflict diamonds is to smuggle them to the markets of countries who have agreed to the Kimberley process and then to mix them there with local 'innocent' diamonds.

From various evaluation reports is appears that to be sure, monitoring does exist in the conflict countries, but not in neighbouring countries such as Tanzania and the Central African Republic or Guinea. This is why it is precisely through these countries that diamonds are smuggled to a processing country such as Belgium, or to a country which is not a member of the Kimberley Process (*Diamond Industry Annual Report*, 2004). Criminal organizations take advantage of the weakest elements of the monitoring system to smuggle diamonds. One of the most important destinations for the smugglers is Antwerp.

Antwerp and Conflict Diamonds

As early as the fourteenth century, Antwerp was regarded as the capital of the diamond world, as the largest international trading market and centre for diamond processing. The diamond sector has always been one of the most important economic sectors in Belgium. There are four diamond stock exchanges, with more than 4,500 independent diamond dealers. In addition, thousands of dealers work outside the four official stock exchanges. In various sources it is stated that between

70 and 90% of the trade in rough diamonds passes through Antwerp (Belgische Senaat, 1998–1999, 2001, 2002).

The heart of the diamond world is located in the Diamond Quarter, the 'Square Mile', which consists of seven streets close to the Central Station. In the media, there is often a connection made between the 'official' diamond area and the 'criminal' Pelikaanstraat, a sort of bazaar of gold shops, where Russians and Georgians have their businesses (Siegel, 2002: 345–346). Antwerp diamond dealers do their utmost not to be associated with these gold shops which they regard as mala fide (ibid: 347).

Just like many other European cities, Antwerp has a good infrastructure, with an international port and an airport in the vicinity. The geographical location is favourable. Antwerp is in the centre of Europe, the borders with other countries are open and there are good logistical facilities. There is also a rather pleasant social climate, it is inhabited by a multicultural population and there is the same cosmopolitan atmosphere as in many other European cities. What really makes Antwerp different from, for example, Amsterdam or Berlin, is its reputation with tourists and diamond dealers from all corners of the world as 'the capital of the diamond world'. Antwerp Provincial Council is therefore occupied with promoting and strengthening this international image. It is clear that the diamond sector is a trump card, not only for the city of Antwerp, but also for the Belgian economy.

Diamond dealers came to Antwerp for specific reasons, such as the nearness of the De Beers, with their headquarters in London, but also because of 'the fact that Belgium is a relatively safe country for diamond dealers and there are easy connections with Africa' (Belgische Senaat, 2001). Belgium's colonial past and the intensive contact with the Democratic Republic of Congo, one of the important producers of rough diamonds, also play a role in this. Moreover, according to the HRD, Antwerp offers all the essential facilities: banks and insurance companies who specialise in the diamond trade, diamond stock exchanges, special regulations regarding taxes and inspection and the fact that the city functions as a world centre for sorting diamonds (ibid, 2002).

These circumstances are, however, not only attractive for bona fide diamond traders. Except for the 'official' diamond dealers who are members of one of the four diamond stock exchanges in Antwerp, there are also innumerable intermediaries, brokers and other dealers. 'It is precisely with these unregulated dealers that the danger of criminal organizations can be found' (ibid). Antwerp plays a role as meeting place for divergent criminal entrepreneurs. Foreign criminals come to Antwerp to meet in relative anonymity and make agreements with each other about planning future activities. A number of them are occupied with the illegal import and transit of diamonds from areas of conflict.

In a report by the Belgian Military Intelligence, it is stated that there are networks existing in Antwerp which are occupied with smuggling diamonds from Angola. The names of various Belgian diamond traders and diamond firms are mentioned in this report (cited in Vander Beken et al., 2004: 275). Some diamond companies supply false declarations of origin, for example from Zambia, a country without conflicts where absolutely no diamonds can be found (Security Council, 2002). There are also known cases of diamond dealers who, 'outside the normal circuit on an individual basis via couriers, have diamonds imported directly from Africa for their own use and exploitation' (Belgische Senaat, 2000).

In principle, in Belgium there is a well-organized system which makes it possible to legally import diamonds in a supple manner. 'No single diamond dealer who knows this system would take the risk to smuggle diamonds into Belgium. Those who do are mostly amateurs and people from Africa who come to Belgium for the first time with goods that are still readily available over there' (Belgische Senaat, 1998–1999). There are regular reports in the media about arrests of smugglers at customs controls. They enter Europe via Zaventem airport. Many are members of a larger network and are transporting diamonds which they have been given by their clients in Africa. In Belguim, they do not declare possession of the diamonds at the customs. Diamonds are hidden in the luggage, or stuck to the body or swallowed. The smugglers often come to Belgium with a tourist visa for 3 months. In this period they try to sell the diamonds on the parallel market (Vander Beken et al., 2004: 279). This parallel market is considered to be one of the three Antwerp diamond markets: the white, the grey and the black (Smile et al., 2000). The 'white market' concerns the legal and official purchasing and selling of diamonds at the diamond stock exchanges; the 'grey market' is the unofficial, independent and parallel market consisting of unregistered and therefore unofficial jewellers, brokers and sellers; on the so-called 'black market' everything happens which is considered illegal, here smugglers and dealers operate in illegal goods. The borders between the three markets are not always clear and according to some authors, the distinction is artificial (Smile et al., 2000: 29–31). During my fieldwork in Antwerp, I found that there was contact between official and unofficial diamond dealers and brokers. The impenetrable and secretive world of the diamond trade is, despite its reputation, not, however, completely hermetically sealed. Because of this, smuggled diamonds which are sold on the 'grey market' can still come on to the legal market via mediators or brokers. The Belgian police have also observed that the line between dubious entrepreneurs and respectable diamond dealers is sometimes transgressed (information presented in Vander Beken et al., 2004: 286).

The difference between the three markets is, however, exaggerated. In my opinion, three clearly defined diamond markets do not exist, but rather one big one, in which the cooperation between legal and illegal players varies, depending on the situation. In some cases this cooperation is symbiotic, both legal and illegal dealers benefit from it. Paoli's notions that the supply of illicit goods, in this case 'conflict diamonds', is disorganized rather than strictly planned, and that different individuals operate in the context of flexible networks, can be confirmed here.

The official organizations of diamond dealers stress that they played an active role in the Kimberley Process and are in complete cooperation with all the international authorities concerned. In particular, the 'certificate of origin' is a trump card. The emphasis on the use of certificates has the goal of excluding blood diamonds from the legal sector, but according to the experts, the certificates are ineffective. 'The certificate will not contribute to solving the problem' (Belgische Senaat,

2001). They are at the very most effective in the sense that the illusion is created that the consumer is buying a diamond which is untainted with blood.

Organized Crime: How to Steal a Diamond?

Apart from the dealers and smugglers who operate on the dubious markets, organized crime is also frequently mentioned (Belgische Senaat, 2001; Peleman, 2002; Siegel, 2005). Criminal organizations also use the infrastructure of Antwerp's legitimate world for their business. The social environment, the presence of partners and contacts to conceal criminal activities, is important in this (Kleemans, van den Berg, & van de Bunt, 1998). 'The diamond trade is very vulnerable to infiltration by organized crime. More than that, infiltration is even a fact. Conflict economies attract organized crime like a magnet' (Belgische Senaat, 2001).

In the 1990s, many top Russian top criminals lived in Antwerp and ran their companies from there. The link between these businesses and smuggling diamonds from areas of conflict has not been a secret for some time now. In the second half of the 1990s, the Russian mafia had operations in Sierra Leone and used its firms in Antwerp as a cover for illegal activities (Peleman, 1999). There is further evidence that Russian and Ukrainian arms dealers are directly involved in the diamond trade in Antwerp (Belgische Senaat, 2001).

The geographical nearness of the Pelikaanstraat, with its Georgian and Russian gold shops, is not favourable for the image of the official Diamond Quarter. The diamond traders accuse their neighbours of tarnishing the good name of the Antwerp diamond world and of carrying out criminal activities (Siegel, 2002).

The activities of criminal organizations are not only restricted to smuggling. Spectacular robberies from museums, airports and even from the 'heart' of the diamond headquarters take place regularly in the capital of the diamond world. Antwerp's Diamond Centre thus became the site of a $118 million robbery on 16 February 2003. One hundred and twenty-three safes at the Grunberger diamond office building were robbed. Investigators have tried to piece together how the thieves managed to break into one of the world's most secure vaults, drawing the conclusion that it could only have happened with assistance from inside the diamond dealers' community. One of the main arguments was that the thieves were able to avoid security in the building and the door was not forced; later it appeared that they copied keys and replaced video cassettes. Allegedly they were in possession of security passes and knew the exact position of the security cameras. Despite all these cameras, alarms, 24-h security guards on every entrance and exit, and cameras in the vault, the robbery was not prevented.[2]

---

[2] The Guardian, February 19, 2003; Rapaport News, February 17, 2003: http://www.diamonds. net/news; CNN. com, February 28, 2003: http://edition.cnn.com/2003/WORLD?europe/02/27/ belgium.diamonds.ap/

The robbery is considered to have been carried out by Italian organized crime, which has been dubbed in the media as the School of Turin.[3] No evidence of the existence of such a school, however, was found during my research, which brought me to the conclusion that this 'gentlemen Mafiosi' school is a legend rather than an actual group. A few months before the robbery, Leonardo Notarbartolo from Trana, a suburb of Turin, rented an office in the Grunberger Building. Strangely enough he was never screened as the other diamond dealers are. Neither was his firm Damaros officially registered, nor were any import or export activities observed. Notarbartolo showed up in Antwerp once every 3 months, but 5 days before the robbery he entered the building daily. Another suspect, Antonio Faletti, was allegedly the brain of the operation.

According to the police, the professionalism of the thieves was evident from the way in which the alarm system was manipulated: it appeared to function normally, but nothing was registered. The video cassettes had apparently been replaced by blank tapes. The selection of the stones was also well-planned: rough or half-cut stones, which are more difficult to sell, were not taken. Though the thieves were arrested, none of the stolen stones were ever recovered.

Only a few months later, on Sunday morning, 30 November, another robbery took place, this time from Antwerp's Diamond Museum, which was holding a temporary exhibition 'Art Deco – diamond jewels 1920–1939 – Glamorous gems for the woman of the world'. This time thieves entered as regular visitors, overrode the alarm system, smashed two display cases with sledgehammers and took classic diamond jewellery consisting of many small diamonds, which are easier to sell.[4]

European international and national airports are also the traditional place for professional robberies to take place. Almost all the attacks take place during the transportation of the goods at the time of landing, take-offs, or during unloading. In the summer of 2001, the robberies which took place at Brussels international Airport led to twelve European airlines refusing to transport diamonds and money to the airport. A Sabena airplane was attacked by masked and armed robbers who stole an enormous amount of jewellery and money from the parked plane in 2000. One year later, robbers escaped with diamonds and other valuable goods worth $160 million from a Lufthansa jet cargo. At the moment of the attack the staff were preparing to transfer the boxes of diamonds, valued at 6.5 million euro, to a Lufthansa flight from Brussels to Frankfurt.[5] From the police investigation it became clear that a number of airport personnel were involved in robberies from post packages from flights to Zurich, Geneva, Madrid and Stockholm.[6]

---

[3] BBC News, February 14, 2004: http://news.bbc.co.uk/go/pr/fr/-/1/hi/world/europe/3364911.stm

[4] Rapaport News, December 5, 2003: http://www.diamonds.net/news/; Gems and Jewellery. Trade news, 2003; http://www.gemsgateway.com/trade_news/trade_news_3html

[5] Jewelers Circular Keystone, July 9, 2001; Daily Times, June 15, 2002: http://www.dailytimes.com.pk/default.asp?page = story_15–06–2002_pg4_10

[6] The stolen diamonds were later sold on the black market in Namen and Charleroi, in: *De Morgen*, September 7, 2002, *Gazet van Antwerpen*, September 9, 2002. In other European airports robberies of diamond transports also take place, such as Roissy in Paris in August 2002. In 2005, the most spectacular robbery took place at Schiphol Airport, Amsterdam, which was carried out by a Dutch criminal group, according to the police.

These large-scale robberies of diamonds are examples of the professionalism of the criminal groups, who do not need assistance from inside the diamond sector or the airport personnel. This can be directly associated with experienced and well-equipped criminal organizations. On the other hand, Paoli's second paradox relating to mafia-groups which are not exclusively involved in activities in illegal markets, can be confirmed. Diamonds are not illicit products, but their theft results in enormous profits for large-scale criminal organizations.

Conclusions

In this article, I have analysed the involvement of criminal groups in diamond-related crimes. Two specific activities of organized crime in Antwerp were discussed: the smuggling of so-called conflict diamonds and diamond robberies.

As of the end of the 1990s, conflict diamonds have been high on the political agenda. The Kimberley agreements, which have been in effect since 1 January 2003, were entered into to provide a solution for the problem. Yet smuggling from areas of conflict into Europe remains difficult to combat. Antwerp offers considerable facilities for bona fide diamond dealers, but also for smugglers and criminals. In the city, meetings take place between criminals and agreements are made about criminal activities. In the so-called 'grey' or parallel market, trading takes place of smuggled diamonds from Africa and sometimes they end up on the legal market through contact with 'official' diamond dealers. Moreover, the social environment in Antwerp is important for the activities of organized crime. Antwerp is by necessity the end point of the smuggling, because it is here that rough diamonds are traded. The Antwerp 'infrastructure' of diamond dealing and processing constitutes an essential part of the chain. Such a situation does not exist in smuggling methods where a final product is supplied.

Organized crime also manifests itself in spectacular robberies of museums, in airports and even in the centre of the Diamond Quarter itself. Either with the assistance from inside the diamond community or from the outside, the interest of organized crime in diamonds is enormous. The stolen stones are easy to smuggle, they are usually not recovered and the profits are huge.

In this article, I have considered whether organized crime was behind the smuggling activities of conflict diamonds to Antwerp on the one hand and of the numerous robberies on the other. Referring to Paoli's concern with the concept of organized crime, I have found that whereas for the smuggling activities of conflict diamonds, individual and small-scale networks of African smugglers were making deals with some legal dealers, the spectacular and well-organized robberies were probably conducted by professional criminal groups. The word 'probably' points to the sophisticated character of the crimes, since in most of the cases the offenders were not caught (except for the Turin criminals) and in no cases were the stolen diamonds recovered.

It is not only the criminal groups who benefit from diamond-related crime. In the beginning of this article I argued that various players involved in the regulation and control of the diamond industry, such as NGOs and diamond organizations, profited from the attention that the scandals concerning conflict diamonds and sensational robberies received. The construction of the 'mysterious attractive power' of diamonds as an object of crime, is perhaps the best advertisement for this luxury product on both the legal and illegal markets.

References

Bauman, Z. (1998). *Globalization: The human consequences*. New York: Colombia University Press.

Beck, U. (2000). *What is globalization?* Cambridge: Polity Press.

Belgische Senaat. (1998). Parlementaire Commissie van Onderzoek naar de Georganiseerde Criminaliteit in België, *Eindverslag*, zitting 1998–1999, nr. 1–326/9.

Belgische Senaat. (2000). *De Georganiseerde criminaliteit in België; eerste tussentijds verslag namens de Opvolgingscommissies inzake de Georganiseerde criminaliteit*, zitting 2000–2001, nr. 2–425/1.

Belgische Senaat. (2001). Parlementaire onderzoekscommissie 'Grote Meren', Hoorzittingen, *Verslag*, vrijdag 7 december, gr. 4.

Belgische Senaat. (2002). Parlementaire onderzoekscommissie 'Grote Meren', Hoorzittingen, *Verslag*, vrijdag 11 januari, gr. 7.

Bossard, A. (1990). *Transnational crime and criminal law*. Chicago: Office of International Criminal Justice.

Eriksen, Th. (2003). Introduction. In Eriksen, Th. (Ed.), *Globalization. Studies in anthropology*. London, Sterling Virginia: Pluto Press.

Fijnaut, C., & Paoli, L. (2004). General introduction. In C. Fijnaut & L. Paoli (Eds.) *Organized crime in Europe*. Dordrecht: Springer.

Galeotti, M. (2004). Introduction. Global crime today. *Global Crime, 6*(1), 1–7.

GAO. (2002). *International trade. Critical issues remain in deterring conflict diamond trade*. Report to Congressional Requesters. June.

Global Witness. (1998). *Rough trade: The role of companies and governments in the Angolan conflict*. December.

Kleemans, E., van den Berg, E. A. I. M., & van de Bunt, H. G. (1998). *Georganiseerde criminaliteit in Nederland – Rapportage op basis van de WODC-monitor*. Onderzoek en beleid nr. 173. Den Haag: WODC.

Mittelman, J. (2000). *The globalization syndrome; transformations and resistance*. Princeton, NJ: Princeton University Press.

National Geographic March 2002, pp. 1–51.

Paoli, L. (2002). The paradoxes of organized crime. *Crime, Law and Social Change, 37*, 51–97.

Passas, N. (2002). Cross-border crime and the interface between legal and illegal actors. In P. Van Duyne, K. von Lampe, & N. Passas (Eds.) *Upperworld and underworld in cross-border crime*. Nijmegen: Wolf Legal Publishers.

Peleman, J. (1999). *Sierra Leone en de diamant-huurlingen*. Antwerpen: IPIS brochures, nr. 114.

Peleman, J. (2002). Conflictdiamanten – gezien vanuit Afrika. *Tijdschrift voor Criminologie, 44*(4), 358–366.

Ruud, A. (2000). Corruption as everyday practice. The public-private divide in local Indian society. *Forum for Development Studies, 2*, 271–294.

Security Council. (2002). *2002/486. Additional report of the monitoring mechanism on sanctions against UNITA*, 26 April.

Shore, C., & Haller, D. (2005). *Corruption. Anthropological perspective*. London: Pluto Press (nog te verschijnen).

Siegel, D. (2002). De joodse gemeenschap en de Antwerpse diamantsector in historisch perspectief. *Tijdschrift voor Criminologie, 44*(4), 338–350.

Siegel, D. (2005). *Russische bizniz.* Amsterdam: Meulenhoff.

Smile, I., Gberie, L., & Hazleton, R. (2000). *The heart of the matter: Sierra Leone, diamonds and human security.* Partnership Africa Canada, January.

Tamm, I. J. (2002). *Diamonds in peace and war: Severing the conflict-diamond connection.* World Peace Foundation: WPF Reports, nr. 30.

Vander Beken, T., Cuyvers, L., De Ruyver, B., Defruytier, M., & Hansens, J. (2004). *Kwetsbaarheid voor georganiseerde criminaliteit. Een gevalstudie van de diamantsector.* Gent: Academia Press.

Weyzig, F. (2004). The Kimberly Process Certification Scheme one year after: State of affairs in the European Union. Fatal Transactions, SOMO, NIZA.

Chapter 8
Eco-Crime: The Tropical Timber Trade

Tim Boekhout van Solinge

Introduction

In April 2002, during the UN Convention on Biological Diversity (BCD) in The Hague, environmental groups held actions in several European countries: Germany, Finland, France, Italy, and the Netherlands. A cargo ship was boarded off the Dutch coast by Greenpeace activists who chained themselves to the ship, climbed the mast and displayed a banner saying "Europe, stop ancient forest destruction."[1] The ship transported timber of the Oriental Timber Company (OTC), the largest logging company in Liberia, West Africa. The next day, Greenpeace activists climbed the cranes in the Amsterdam timber harbour, preventing the Liberian timber from being unloaded. On the third day, the police arrested the activists, who were later released without charge. The environmental groups Greenpeace and Global Witness claimed that the Liberian timber on board came from a unique ancient forest in West Africa. The NGOs also claimed the timber trade was connected to arms trafficking. The timber companies concerned reiterated that their timber cargo was legal, as it had been logged and exported with the authorisation of the Liberian authorities, a fact conceded by the environmental groups.

The Liberian timber targeted by activists is a good example of "conflict timber." Conflict timber refers to timber trade that is related to armed conflicts. The term was first coined in 2001 by a UN panel of experts investigating the illegal exploitation of natural resources in the Democratic Republic of Congo (DRC). The fact that timber is used to fund armed conflicts does not mean conflict timber is an illegal commodity. Although this type of trade is often legal, it can be argued that it is "dirty" or contaminated in other ways, which could justify it not being allowed and be declared illegal. NGO Global Witness described many examples – Cambodia, Liberia, Burma, Zimbabwe, the Democratic Republic of Congo and Cameroon – of how little and how slow the international community reacts to the trade in *Logs of War* (Global Witness, 2002).

---

[1] In 2001, in the same Amsterdam timber harbour, Greenpeace activists had put red paint on OTC timber, which represented the bleeding to death of ancient forests.

D. Siegel and H. Nelen (eds.), *Organized Crime: Culture, Markets and Policies.*
© Springer 2008

Conflict timber should be distinguished from "illegal timber," another category of timber trade, which is also subject to actions by environmental groups. Illegal timber refers, in most cases, to timber that has been logged illegally, for example without permits or in protected forests. In recent years, environmental and human rights groups have accused companies and governments of being involved or complicit in illegal or unethical logging and timber trade activities. NGOs present themselves as the vanguards of morality, laws, and regulations, accusing companies and governments of being weak law enforcers or irresponsible law-breakers. NGOs sometimes take the prosecutor's role, by doing – sometimes risky – investigative research in order to be able to put individuals or organisations in the position of the accused.[2]

A substantial part of the timber trade, in tropical timber in particular, concerns illegal timber. It is a common estimate that about half of the tropical timber on the European and western markets is still of illegal origin (Friends of the Earth, 2001; Jaakko Pöyry Consulting, 2005). Until now, criminologists have generally shown little interest in the trade in (tropical) timber and other wildlife. As Nigel South noted, the plundering of the earth's natural resources has not been thought of as a crime until recently. The earth and its resources are being wasted and overexploited, a practice in which numerous crimes, violations, deviations and irregularities are perpetrated against the environment. South therefore proposes to label them as "green crimes," which he generally defines as crimes against the environment (cf. Carrabine et al., 2004). In a recent book, *Issues in Green Criminology*, Beirne and South (2007) developed this concept further. In the 1990s, Gray proposed the concept of the international delict "ecocide," which he defined as "causing or permitting harm to the natural environment on a massive scale" which would "breach a duty of care owed to humanity in general" (Gray, 1996). Despite the existence of these publications, lawyers and criminologists are, generally speaking, largely neglecting the profitable and often illegal trade in tropical timber and other wildlife such as protected animals. There is, however, every reason to question the small amount of attention this type of trade gets, considering its repercussions. Natural habitats are currently disappearing at such a speed that a new wave of mass extinction of species might occur in the twenty-first century (cf. Leakey & Lewin, 1996).

To better address the loss of species and the extinction of the world's ecology as a result of illegal activities it may be useful to use more precise language. As ecology refers to the study of the relationship between plants, animals, people, and their environment,[3] ecological crime – abbreviated eco-crime – could be defined as the (illegal) procurement and trade in living elements of ecosystems, which may pose a threat to that (unique) ecosystem.[4] The expression eco-crime has some overlap with the illegal wildlife trade and the more generic terms of green or environmental

---

[2] The Environmental Investigation Agency (EIA), for example, does undercover research into the illegal trade in rare and endangered species: ivory, tiger and leopard skins, and timber.

[3] Definition taken from Collins English Dictionary for Advanced Learners (2001).

[4] An ecosystem is the physical environment plus the organisms living in it of a particular habitat, such as a forest or a coral reef. See Wilson (2002), *The future of life*, 214.

crimes, but "eco-crime" refers more directly to illegal activities concerning elements of ecosystems, namely flora and fauna. Eco-crime thus includes the illegal timber trade, the illegal wildlife trade (birds, big cats, apes, snakes, etc.) as well as illegally trading wildlife products such as ivory and timber. Conflict timber can also be considered to be a form of eco-crime. Although technically not always illegal, this type of trade is often connected to other illegal activities, as the Liberian timber trade described in the next section will illustrate.

This article discusses one type of eco-crime in particular: the large-scale logging of the world's tropical forests. It shows that an everyday and age-old commodity such as timber is the subject of organized crime activities. The opportunities and profitability of the tropical timber trade are generally underestimated and neglected. Illegal entrepreneurs and international crime syndicates, sometimes with the involvement of armed forces and governments, have become involved in the international timber industry. It makes the timber industry a good example of how legal and illegal economies are intertwined and overlap each other. Two types of timber trade will be discussed: conflict timber and illegal timber.

Conflict Timber: Taylor-Made Hardwood

From the moment former rebel leader Charles Taylor was elected President of the West African country of Liberia in 1997, he used different sectors of the economy in order to buy weaponry. Firstly, he used the large revenues of ship registrations. Being a "flag of convenience" tax haven, Liberia has the world's second largest registered commercial fleet (after Panama). About one third of all oil tankers, cruise ships and freight carriers are registered as Liberian. The ship registration represented 30% of Liberia's national budget.

A second important source of income for President Charles Taylor's regime was diamonds. Liberia was exporting quantities of diamonds exceeding the country's potential. Many diamonds actually came from neighbouring Sierra Leone, from mines controlled by rebels of the Revolutionary United Front (RUF). Diamonds enabled the RUF to buy weapons, which were smuggled in from neighbouring Liberia. Liberia's President Charles Taylor was the facilitator of the arms for diamonds trade, just like he had been involved in several other West African conflicts, presumably with the objective of capturing some of their valuable resources, particularly diamonds and gold (cf. Baker et al., 2003). In 2000, a report to the UN Security Council clearly described how Liberia had become a trans-shipment platform for arms to the RUF rebels, with President Charles Taylor "and a small coterie of officials and private businessmen" being involved in covert operations such as diamond smuggling and the extraction of natural resources in Liberia and other countries. "Paymaster" of the trade was Talal El-Ndine, a wealthy Lebanese businessman. "Liberians fighting in Sierra Leone alongside the RUF, and those bringing diamonds out of Sierra Leone are paid by him personally" (UN Security Council, 2000: 37). In 2001, the UN extended its already existing arms ban against Liberia (after the 1989–1996 war) to diamonds, in order to prevent diamonds from Sierra Leone from being exported through Liberia. The diamond trade being embargoed, Taylor had to look for other sources of income.

Liberia was the only country in West Africa still having a significant portion of its original rainforest cover. Under Taylor's regime, large quantities of Liberian timber and especially the valuable hard wood were exploited and exported. It soon became clear that Liberia's timber industry was related to arms trafficking. The UN's Expert Panel Report on Sierra Leone made no doubt about it. "The principals in Liberia's timber industry are involved in a variety of illicit activities, and large amounts of the proceeds are used to pay for extra-budgetary activities, including the acquisition of weapons." One of the members of "Taylor's coterie," as the Report described it, was Simon Rosenblum, an Israeli based in Abidjan, Ivory Coast, carrying a Liberian diplomatic passport. "He has logging and road construction interests in Liberia and his trucks have been used to carry weapons from Robertfield, Liberia, to the border with Sierra Leone." The UN Report described how weapons were flown in by a BAC-111, owned by Leonid Minin, an Israeli businessman of Ukrainian origin and known arms trafficker (UN Security Council, 2000: 36–37).

> Minin was, and may remain, a confidant and business partner of Liberian President Charles Taylor. He is identified in the police records of several countries and has a history of involvement in criminal activities ranging from east European organized crime, trafficking in stolen works of art, illegal possession of arms, arms trafficking and money laundering. Minin uses several aliases. He has been refused entry in many countries, including Ukraine, and travels with many different passports. Minin offered the aircraft mentioned above for sale to Taylor, as a Presidential jet, and for a period between 1998 and 1999 it was used for this purpose. It was also used to transport arms (UN Security Council, 2000: 36).

In Liberia, Minin was the owner of the Exotic Tropical Timber Enterprise (ETTE). In the UN Expert Panel Report on Sierra Leone he is mentioned as one of the two individuals of Taylor's circle who are particularly connected to the timber trade. The other one was Guus van Kouwenhoven, a Dutchman who started a hotel and gambling business in Liberia in the 1980s.[5] He would also use Liberia as a stopover for hashish transports from Pakistan to the Netherlands. Dutch police investigations into his activities were stopped because the cooperation with the Liberian authorities did not proceed smoothly (Ziel, 2003). In the late 1990s, Van Kouwenhoven brought the Oriental Timber Company (OTC) from Asia to Liberia, first under the name of the Liberian-Malaysian Timber Company. "Minin and Van Kouwenhoven are the people who are linked to Liberia's timber industry, thus providing a large amount of unrecorded extra-budgetary income to President Taylor for unspecified purposes. Three companies are involved: Exotic Tropical Timber Enterprise (ETTE), Forum Liberia and the Indonesian-owned Oriental Timber company" (UN Security Council, 2000: 37). Van Kouwenhoven was Chairman of the Malaysian-based Oriental Timber

---

[5] It was in his Hotel Africa in the capital Monrovia where the pilots and crew of Minin's aircraft, used for clandestine arms shipments into or out of Liberia, always stayed. See: UN Security Council S/2000/1195, p. 37.

Company (OTC), the biggest logging company in Liberia and the largest single foreign investor in Liberia. He also was managing director of the second largest logging company: Royal Timber Corporation (RTC). "Mr Gus," as he was locally known, played a key role in the Liberian timber trade and weapons smuggling.

> Van Kouwenhoven is responsible for the logistical aspects of many of the arms deals. Through his interests in a Malaysian timber project in Liberia, he organises the transfer of weaponry from Monrovia to Sierra Leone. Roads built and maintained for timber contraction are also conveniently used for weapons movement within Liberia, and for the onward shipment of weapons to Sierra Leone (UN Security Council, 2000: 37).

Van Kouwenhoven was also a member of the Liberian Forestry Development Authority (FDA). Talal El-Ndine, the Lebanese businessman financing the diamond smuggling, was another FDA member. President Charles Taylor's brother, D. Robert Taylor, was the FDA's Managing Director (Beaumont, 2001). The FDA ensured the logging and timber export practices were going according to the rules, and were thus legal.

In December 2000, the UN's Expert Panel on Sierra Leone recommended to the Security Council to "(place) a temporary embargo on Liberian timber exports, until Liberia demonstrates convincingly that it is no longer involved in the trafficking of arms to, or diamonds from, Sierra Leone" (UN Security Council, 2000: 13). In March 2001, Greenpeace Spain published a report on the Liberian timber trade, *Logs of War*, stating Liberia's rainforests were now in the hands of "war lords." It also criticised Spanish-owned timber companies Forum Africa and Forum Liberia for being involved in the timber and arms trade, by doing business with Leonid Minin's timber company ETTE. The Greenpeace report also makes mention of timber for arms trade dating back to the 1990s, when "France supplied arms to the NPFL (Charles Taylor's armed faction) in exchange for timber" (Greenpeace, 2001: 10). Greenpeace subsequently asked for a total boycott of Liberian timber. Timber export had become Liberia's main source of revenue, allowing Taylor to stay in power. Logging in Liberia was occurring at such a fast rate that in 5 years time half of the Liberian rainforest had disappeared. Logging giant OTC, which controlled one third of Liberia's rainforest, obviously had a large share in the rapid exploitation of the rainforest.

In May 2001, the UN Security Council decided not to impose timber sanctions against Liberia. France and China, both permanent members of the UN Security Council and the main importers of Liberian wood, opposed timber's inclusion in the sanctions, arguing there was not enough evidence to link it to regional conflicts (Aloisi, 2001). The UN Security Council did decide to extend the existing arms sanctions to diamonds. It also banned foreign travel by members of Taylor's inner circle, including Minin and Van Kouwenhoven. Journalist Peter Beaumont reported about the UN Security Council decision in *The Observer:* "The trade in timber – to the exasperation of Britain and the United States – was exempted at the insistence of France, which imports up to a third of it" (Beaumont, 2001). "The new UN sanctions regime is utterly pointless," complained a European diplomat to him, "while Taylor is still able to keep exporting timber and bringing in guns." Beaumont went to Liberia where he learned that the timber trade was a "business run with military precision." He was told that timber baron Van Kouwenhoven was Charles Taylor's most important business ally. The port of Buchanan had actually been handed over to OTC, to run as its private city. The 108-mile dirt road from Buchanan to Greenville was upgraded to a four-lane highway, allowing logging to continue everyday of the year. A businessman familiar with Taylor's business told him: "Look, it is an open secret. Gus fronted Taylor up $5 million for his logging concessions. They split the profits. Gus's ships take out the logs and they bring in the guns. It was the same deal with the diamonds" (Beaumont, 2001).

NGOs did not stop lobbying. Later in 2001, Global Witness Director Patrick Alley declared having specific examples of logging ships arriving in Liberia, unloading arms, for example from China, and loading up with logs. "The logging industry clearly has to make a profit, but we think that a major part of that money goes into funding regional conflicts – you can buy a lot of guns with $100 million" (Aloisi, 2001). In March 2002, Global Witness issued a report on conflict timber around the world, *Logs of War*, and asked again for an international boycott of Liberian timber as it was related to arms smuggling (Global Witness, 2002). A month later, during the UN Convention on Biological Diversity (BCD) in The Hague, OTC timber was the target of environmental activists – as described at the start of this article.

It took more than another year, until July 2003, before the UN Security Council decided to introduce timber sanctions against Liberia. Liberian timber could no longer be legally traded. The reasons for the timber sanction were not ecological – protecting Liberia's rainforest – but political and military: the Liberian timber trade network was used for smuggling weaponry, which seriously hampered the UN peace process in Sierra Leone, where the UN had stationed its largest peacekeeping force.

In August 2003, 1 month after the timber sanctions, Charles Taylor resigned after international pressure. It allowed the signing of a peace treaty between the government and the rebels, ending a 14-year civil war. Soon after, an international warrant was issued for Taylor's arrest. The Special Court for Sierra Leone (SCSL) charged him with war crimes and crimes against humanity, for having created and backed the RUF rebels during the Sierra Leone war. Taylor first found asylum in Nigeria, but in 2006, Nigerian authorities put him on a plane to Liberia. He was then handed to the UN Special Court for Sierra Leone (SCSL), which will conduct the trial's proceedings, but the trial itself takes place at the International Criminal Court in The Hague.

Leonid Minin was arrested in Italy in 2001, in room 341 of Hotel Europa near Milan, which he owned himself, in the company of four prostitutes. Twenty grams of cocaine were found as well, and $150,000 in cash and more than half a million in diamonds. There was also a cache of 1,500 documents detailing Minin's dealings in oil, timber, gems and guns. Minin was convicted to 2 years in prison for the drug offence (Traynor, 2001). Italian prosecutors then accused him of dealing in conflict diamonds and timber, but the Italian court found it had no jurisdiction, since the illegal shipments did not pass through Italy. Minin was only fined ($51,000) for illegally possessing the diamonds. Global Witness said it is hoping to reopen the case against Minin with new documents.

Van Kouwenhoven managed to leave Liberia, escaping the international travel ban in place since 2001. In October 2002, the multimillionaire timber trader showed up as a "new rich" in Dutch daily *NRC Handelsblad* and financial monthly *Quote*. He had entered the ranks of the 500 most wealthy Dutch.[6] In 2004, Greenpeace uncovered that he was in Congo-Brazzaville, involved with logging company Afribois.[7] In 2005, Van Kouwenhoven was arrested in his hometown Rotterdam. A year later he was convicted to 8 years in prison by a Dutch court. He was found guilty of being in breach of a United Nations arms embargo on Liberia.[8] The court found that he had traded guns for timber and used his timber business to smuggle weapons used by militias, committing atrocities against civilians in West Africa.

Illegal Tropical Timber

Dutch timber baron Van Kouwenhoven was not convicted for his logging activities. As he logged legally, Liberian timber could be exported world-wide. The millions Van Kouwenhoven made, are exemplary of how lucrative the timber trade is. Especially tropical timber is very valuable: species are priced hundreds and sometimes thousands of euros per cubic meter. The Liberian case also shows the lack of international monitoring and control of the opaque timber trade. Conflict timber does not get the same public attention as conflict or blood diamonds, as they are better known.

While the timber from Liberia involved conflict timber with a legal status, much tropical timber on the international markets is clearly *illegal*. The illegality of timber can take several forms. The most evident but not most common example of illegal timber concerns timber listed by the Convention on the International Trade in Endangered Species (CITES). Depending on the appendix a particular animal or plant species is listed on, the trade in CITES Species requires special permits. It is the task of the customs and national CITES bureau to oversee whether the import of protected CITES species – from timber to amphibians and tigers – is proceeding correctly.

A more prevalent type of illegal timber concerns timber that was logged illegally. Debra J. Callister gave an overview of what may include illegal logging: logging protected species, logging outside concession boundaries, removal of oversized or undersized trees, extraction of more than the allowed harvest, harvesting in areas where extraction is prohibited such as steep river slopes, river banks and catchment

---

[6] The travel ban was introduced by the UN Security Council in May 2001 to Taylor's inner circle: timber traders Leonid Minin and Guus van Kouwenhoven, and arms dealers Victor Bout and Sanjivan Ruprah. Van Kouwenhoven was 370 in the "Quote 500," made by Dutch financial monthly *Quote*. See Hooghiemstra (2002), In Gesprek over de nieuwe rijken met Jort Kelder, hoofdredacteur van Quote.

[7] Greenpeace claimed the Danzer group continued to finance Kouwenhoven's timber trade. See Greenpeace (2005), *Danzer Group involved in bribery, illegal logging, dealings with blacklisted arms trafficker and suspected of forgery*.

[8] Global Witness provided some of the evidence in the case against Van Kouwenhoven.

areas. It also covers unauthorised logging, such as logging without a permit or license or in breach of the terms of the license. Other illegal activities related to logging include: timber smuggling, transfer pricing, under-grading, under-measuring, under valuing and the misclassification of species (Callister, 1992: 6).

At least half of the tropical timber on the international market is thought to be of illegal origin.[9] In practice, the illegal timber from South America, Sub Sahara Africa and South East Asia can easily be imported into the major destination countries: Japan, EU, USA and, increasingly, China. The timber then ends up in construction works or at retailing shops. When customers buy attractive hardwood (products) such as floors, doors, garden chairs (or window frames), they are usually unaware it may concern illegal timber. A legal discrepancy occurs here: timber may be illegal according to Brazilian or Indonesian laws, but the illegality of a particular timber transport is hard to prove after it has arrived in Europe or Japan. Customs officials usually do not check more than the paper work, and only occasionally examine whether or not the timber on paper corresponds to the timber in cargo. But they do not always have the knowledge to identify timber types, nor is it always clear which governmental agency should actually do it.[10] As few trade barriers exist in practice, illegally logged timber can easily be traded internationally, which leads some environmentalists to speak of a practice of "green washing."

In the South American Amazon, the world's largest forest, containing 20% of all fresh water, an area the size of Belgium or Switzerland is being logged annually.[11] The estimated share of illegal logging in the Brazilian timber exports is 20–80%. The higher estimates generally refer to illegal logging in Amazonia. The "timber mafia," professionally equipped with chainsaws, bulldozers and guns, roam the Amazonian forests in search of valuable species that can be sold. The harbour of Santarém, in the Brazilian Amazon, is one of the known harbours from where illegal timber is exported. In this harbour, as well as in the Amazonian harbour of Belem, 50–70% of the leaving merchandise concerns timber. In the harbour of Santarém, visited by this author in 2003, the timber was stacked up high on the docks, waiting to be loaded on board by big cranes, the destination painted on the side. The timber is mostly destined for European harbours (in France, Spain, UK, etc.), the Dutch port Flushing (Vlissingen) being the most prevalent.[12]

While the West African rainforests have mostly been logged, more central African countries like Cameroon, the Democratic Republic of Congo and Congo-Brazzaville are currently popular source countries for timber. The reported estimates

---

[9] Several overviews exist, such as Jaakko Pöyry Consulting (2005), *Overview of illegal logging*, based on the overview of the American Forest and Paper Association – AF&PA (2004), *Illegal logging and global wood markets: The competitive impacts on the U.S. wood products industry.*

[10] Such as in the Netherlands, where criminology students discovered that it is not always clear whether the responsibility lies with the customs or with the General Inspection Service (AID), part of the Agricultural Ministry. Both agencies merely point to each other, as they learned after a visit to the port of Flushing.

[11] Large-scale logging has now caused the first draughts in the Amazon, leading to large fires.

[12] As appeared during a visit in September 2003, after arriving by boat in the harbour.

for the share of illegal forest production vary from 34–60% in Ghana, to 50% in Cameroon, and 50–70% in Gabon (Jaakko Pöyry Consulting, 2005). A problem particularly related to logging in Africa is that it further stimulates poaching and the trade in "bush meat," which now seriously threatens some species with extinction. A logging road means a major incursion into a forest, which is usually difficult to access. It enables hunters to go further into the forest and to reach distant markets. They can take their catch to the city by car, or sell it somewhere along the road. As a consequence, "bush meat" is now increasingly featured on the menu of the urban population as well. Conservationists say illegal commercial hunting of African wildlife for sale as "bush meat" has reached alarming levels and immediate action is needed to address the problem before it is too late. Bush meat varies from deer, gorillas and chimpanzees to crocodiles and elephants (Frank, 2001). Much older, of course, is the still thriving illegal trade in ivory, which mostly ends up in Asia.[13]

In South East Asia, most of the Philippines and Thailand has already been logged. Large-scale logging is currently taking place in Indonesia, Malaysia, Myanmar (Burma), Cambodia, Laos and Vietnam. A common method is that timber is logged illegally in one country, and then smuggled to a neighbouring country, from where it is exported. For example, illegal timber from Cambodia is transported to Thailand and Vietnam, where it is manufactured into lawn and garden furniture, which is then exported to Europe, e.g. to Dutch retail chains (Aidenvironment, 2000: 13). Similarly, timber from Myanmar is being (illegally) exported through Thailand. The estimated share of illegal logging in forest production varies from 20–40% in Vietnam, 35% in Malaysia, 50% in Myanmar, 70–80% in Indonesia, 70% in Papua New Guinea to 90% in Cambodia (Jaakko Pöyry Consulting, 2005: 4).

Indonesia is the country where logging is going faster than anywhere else in the world. The large archipelago houses the world's second largest rainforest, with 10% of the world's remaining rainforests. Forests are going down at the speed of six football pitches a minute, which corresponds on a yearly basis to an area larger than Taiwan (Aglionby, 2005; EIA & Telapak, 2004: 4). Most logging, an estimated 70–80%, is illegal. Much more is exported than is officially allowed. The introduction of a policy of decentralisation and regional autonomy, in 2001, made it unclear who has the ultimate right to give out logging permits: the village head (based on customary law), the governors (based on the policy of decentralisation), or the Ministry of Forestry.[14] Another impetus for logging practices is the involvement of the Indonesian army (TNI). As the Indonesian government only supplies 30% of the army's budget, TNI arranges its own revenues through commercial businesses, including a number of logging and timber companies.

---

[13] For an overview see: Naylor (2005), The Underworld of Ivory, *Crime, Law and Social Change*, *42*, 261–295 (2004).

[14] The Center for International Forestry Research (CIFOR), Bogor, Indonesia, has published a series of case studies on the forest impact of decentralisation. See http://www.cifor.cgiar.org.

Logging for timber is actually not the only cause of deforestation. An equally important reason for deforestation is land conversion: turning forests into profitable agricultural land. In Brazil, rich farmers buy land in the Amazonian lowlands and turn it into grass land for cows. With 200 millions cows, Brazil has now surpassed the US as main beef producer. Much of the Amazonian meat ends up in the American meat industry. An even more important reason for deforestation is land conversion for soybean farming. The soy is mostly used as cattle food (for example in the Dutch pigs industry). In Indonesia and Malaysia, rainforests are being converted into oil palm plantations. A lesser known cause of deforestation is the paper and pulp industry. The two largest paper pulp factories in the world are found in Indonesian Sumatra, where Asian paper giants APP and RAPP have been responsible – without being punished – for making paper from virgin rainforests. It threatens the survival of endangered animals such as the forest elephant, rhino and Sumatran tiger.[15]

As a general rule, it regularly occurs that laws are broken during the land conversion practices. For example, the promised social projects and investments are regularly not effectuated, or more is logged than agreed. For the locals, such as forest tribes, it is difficult to address their concerns, as they are dealing with commercially and politically powerful individuals or corporations. In practice, the illegal destruction of some forests not only signifies a threat to the survival of animals and plants living in the forests, but also to the many millions of forest and village people around the world who are directly dependent on forests. It is part of the human experience that people have lived in forests as hunters and gatherers, but this age-old life style is now seriously threatened. In Ivory Coast, Pygmies have disappeared, as the country is "commercially logged out" (Global Witness, 2002: 40). In Kenya, the Ogiek people who have lived since time immemorial in the Mau mountain forest overlooking Kenya's Rift Valley, are currently driven out of the disappearing forests. Not only in Africa, but also in the Amazon and on the large islands of Sumatra, Borneo and Papua, indigenous peoples are being threatened. The voices and rights of forest people, the first human victims of the logging and timber trade, do not easily reach the decision-makers in the economically more developed countries, the destination of the timber. In practice this means that human rights of indigenous forest people are often infringed.

What's at Stake?

In Liberia, half of the unique Liberian rainforest disappeared in a period of 5 years. Liberia's rainforest is considered one of the world's "biological hotspots," an area that is rich in species found nowhere else and environmentally endangered.[16] It

---

[15] As the paper pulp companies did not plant sufficient forest plantations for their paper mills, trees from Sumatra's virgin rainforests (in the province of Riau) have disappeared in paper mills. Indonesian and international NGOs have published about destructive and illegal logging practices of RAPP (part of APRIL) and especially APP (Asian Paper and Pulp).

[16] Definition by biologist Wilson (2002), *The future of life*, 215.

covers almost half of the entire remaining area – and two of the only three large blocks left intact – of the Upper Guinean Forest Ecosystem, a rainforest belt which once stretched over several West African countries. It is home to the only remaining viable populations of the Pygmy hippopotamus, and is the last stronghold of forest elephants in West Africa (Peal, 2000).

Deforestation of the world's tropical forests is reaching alarming levels. The tropical forest may only represent 3% of the earth's surface, but they house more than half of all species on the planet. The speed with which some species are currently disappearing – amphibians for example – is unprecedented (cf. Attenborough, 2000, 2006). Biologists now fear we are at the brink of a new mass extinction.[17] If current trends continue, up to 50% of all species on the planet may go extinct during the twenty-first century. Not only are mammals like panda, tiger, lion, rhino, and apes (chimpanzee, bonobo, orang-utan, gorilla) threatened, so are the wild camel, bears, turtles, dolphins, sharks and many others.[18] In the oceans, large-scale over-fishing threatens species such as cod and tuna. The number of large predator fish in the world's oceans has decreased 90% over the last 50 years. Commonly known fish such as the eel and the shark may not be saved.

The three main reasons for logging in the tropics combined – logging for timber, paper, or land conversion – mean that once giant forests may disappear in only a few years time. For example, if logging in Indonesia continues at the current rate, the forests will have disappeared within 5–10 years. Deforestation in the world and particularly in the tropical rainforests, is going at such a rate that is has become a threat to the planet's biodiversity. The extinction of species is caused by several factors, including global warming, pollution and hunting/fishing. The most important reason for species loss, however, is the destruction of natural habitats, mainly due to human activities. Nowhere is this more apparent than in the tropical forests. It is worth mentioning again that they represent 3% of the earth's surface, but are home to half of all known plant and animal species.

As tropical forests and their inhabitants are disappearing, especially considering the irreversibility of extinction, humanity has to decide soon how many species it would like to preserve on earth. It can easily be argued that it is in the interest of the human species that the world's biodiversity and ecology are not destroyed.[19] Whether this is the case or not, government leaders have actually already decided

---

[17] Mass Extinction Underway, Majority of Biologists Say, *Washington Post*, 21 April 1998. See also Leakey and Lewin (1996), *The sixth extinction. Patterns of life and the future of humankind.* The earth's last mass extinction, the fifth extinction – making the dinosaurs extinct – happened some 65 million years ago. It is assumed that during the last 500 million years, the earth has known five waves of mass extinction.

[18] In 2006, biologists sadly had to conclude that the white Chinese river dolphin had become extinct. Of the white rhino in Africa, which has been around since some 50 million years, only a few are left: four, maybe six.

[19] For example, tropical forests are an important source of water, oxygen, medicine, as well as pleasure (shown by the popularity of jungle tours and safaris, as well as nature documentaries on television).

they wish to keep the world's biodiversity. At the UN Convention on Biological Diversity (BCD) in The Hague 2002, it was decided to "achieve a significant reduction of the current rate of biodiversity loss." Later in the year 2000, at the Johannesburg UN World Summit of Sustainable Development, this target was almost repeated: "the achievement by 2010 of a significant reduction in the current rate of loss of biological diversity." The EU is even more ambitious than the UN. In 2001, during the European Summit in Gothenburg, EU government leaders agreed on the EU strategy for sustainable development, which included that the "biodiversity decline should be halted." For this purpose, the European Commission published a communication on "Halting the loss of biodiversity" by 2010.[20] Although these intentions certainly should be applauded, an EU policy plan does not necessarily mean much in practice, as EU member states eventually have to take most measures (Boekhout van Solinge, 2002). With respect to the loss of biodiversity in Liberia's rainforest, the EU strategy did not prevent Liberian timber from being imported into Europe until the UN timber sanctions of 2003.

If European government leaders are serious – as they should – about halting the loss in biodiversity as they declared in Gothenburg, it is time to think more fundamentally about alternatives to the current economic model of land ownership and the rapid exploitation of the earth resources and species.[21] As far as the tropical forests are concerned, there is currently no (EU) legislation that prohibits the import and marketing of timber or timber products in breach of the laws of the country of origin (Bodard & Pallemaerts, 2005: 64). On the demand side, the consuming countries, it is easily overlooked that their large demand is also driving the illegal logging practices. As long as timber importing countries do not put clearer conditions on their demand for timber, international timber syndicates will continue to flourish.

As a policy response to the illegal timber trade, consuming countries are now opting for legal timber. The EU, US, Japan and the G8 focus on a policy of Forest Law Enforcement, Governance and Trade (FLEGT). As a part of the EU FLEGT action plan, voluntary, bilateral agreements are made with timber producing countries to guarantee legal timber supplies. These agreements are "voluntarily," as they might otherwise conflict with WTO rules of promoting free trade. The definition of "legal timber" refers to the "applicable legal rules in the country of export" (European Commission, 2004: 2). However ambitious the EU Action Plans may seem, they are insufficient to tackle the current rapid disappearance of rainforests. If governments are serious about preserving biological diversity, a more sustainable forestry is the only solution. Sustainable economies however, require radically different sets of economic parameters and indicators to be used in policy making.[22] The FLEGT Action Plan does not choose sustainability, but legality: "In the absence of an international agreement defining sustainable forest management, the focus is

---

[20] See http://www.countdown2010.net.

[21] For a discussion on the economy see Chap. 2 of Wilson (2002), *The future of life*.

[22] For example, a sustainable economy would better take into account energy costs. It would undoubtedly have consequences for some industries such as transport, paper, meat and fish.

pragmatic and offers an opportunity to exclude from the EU market products of some of the most damaging forest practices" (European Commission, 2004: 1). However, as this article has shown, even for "legal" timber such as from Liberia or Malaysia – some of which actually comes from Indonesia – there is no such guarantee. The EU FLEGT plan is therefore not compatible with the European Council Strategy to halt the loss of biodiversity.

Studying Timber Networks

At least half of the tropical timber on the international markets is of illegal origin (Friends of the Earth, 2001; cf. Aidenvironment, 2000; Jaakko Pöyry Consulting, 2005). Although large-scale deforestation also occurs in a country like Russia (particularly in Siberia), from a biological and ecological perspective, deforestation of the tropical forests seems more serious. The tropical timber trade is a highly lucrative multibillion business, in which, as was shown in the case of Liberia, some entrepreneurs manage to make millions. NGOs such as the Environmental Agency (EIA) have shown how international syndicates, often operating from Singapore and Malaysia, are operating large-scale illegal logging operations of Merbau timber from Indonesian Papua to China (EIA & Telapak, 2005, 2006). Criminologists have so far mostly neglected the illegal trade in timber and wildlife.

Illegal logging is obviously strongly related to corruption and, more generally, to weak or failing states (such as Liberia under Taylor). Some source countries such as Indonesia do indict illegal loggers, but the higher echelons of timber trade networks usually escape conviction. The political economy of illegal logging includes (big) logging companies, public officials, workers (sometimes village people from peripheral areas, neglected by governments), law enforcement and the military. Illegal loggers and other wildlife traders will sometimes go far in defending their business, by threatening or murdering environmental activists, as has happened in several countries. Not only do orang-utan protectors in Indonesia need safety precautions, Greenpeace activists in the interior of Brazil are obliged to move around with body guards.

It is often stated in consuming countries, not only by timber traders but also by government officials, that the question of illegal logging should primarily be addressed and resolved in the source countries – where illegal logging takes place. Another argument regularly heard is: "If we do not buy the timber, it will go elsewhere, to China. So what's the point of not buying it? It will be bought anyhow."[23] The argument that the timber will otherwise go elsewhere, is a neutralisation technique – or easy excuse – for continuing to buy illegal timber. An important reason for illegal logging to continue is that the demand side does not insist on buying legal or sustainable

---

[23] The argument was mentioned several times to the author by timber traders, most recently at the 2006 Dutch Timber Fair in Rotterdam.

timber. What is striking about the tropical timber trade, is that many consumers, such as those in European retail shops, are not aware of the fact that much of the tropical timber might be of illegal origin.

Since international regulations and forest law enforcement are generally weak, illegally logged timber can be easily – and cheaply – sold on the international market. And as long as this market is unregulated, illegal entrepreneurs see their opportunities. It enables, for example, Malaysian businessmen to sell illegal timber from national parks in Indonesian Borneo for prices tenfold ($200) of what they pay local loggers for a cubic meter ($20).[24] In the meantime, developing countries have large annual losses as a result of illegal logging (estimated at $10–15 billion in 2002 by The World Bank). The gigantic profits the timber trade can make – sometimes higher than the drugs trade, which also explained why some have moved from drugs to timber – enables bribing, corrupting and involving officials.

Considering that much of the wildlife trade, in both trees and animals, is illegal and destructive, eco-crime is a research area that is worth studying by criminologists and lawyers – next to biologists and conservationists. Criminologists can study the timber trade from the perspective of organized rime, corporate crime, or state crime. They can also propose alternatives, such as ways to make the timber trade more transparent and self-regulatory, to ensure the sustainability of logging and to increase consumer awareness. In line with Gray's proposal of the international delict "ecocide" (Gray, 1996), an interesting legal question is how international and environmental law can contribute to the conservation of biodiversity and some of the world's natural heritage sites.

References

Aglionby, J. (2005). Action plan to save home of new species from loggers, The Guardian, 26 April 2005.

Aidenvironment. (2000). *Herkomst onbekend. Over illegale kap en de Nederlands houtmarkt.* Amsterdam: Aidenvironment.

Aloisi, S. (2001). Liberian timber riches seen fuelling regional war, Reuters Abijan, 9 August 2001.

Attenborough, D. (2000). State of the Planet, three part documentary (DVD), BBC Television.

Attenborough, D. (2006). Planet Earth, thirteen-part documentary (DVD's), BBC Television.

Baker, M., Clausen, R., Kanaan, R., N'Goma, M., Roule, T., & Thomson, J. (2003). Conflict timber: Dimensions of the problem in Asia and Africa. Volume III, African Cases. Burlington, Vermont: ARD Inc. (report for USAID).

Beaumont, P. (2001). How a tyrant's 'logs of war' bring terror to West Africa. The Observer, 27 May 2001.

Beirne, P., & South, N. (Eds.) (2007). *Issues in green criminology. Confronting harms against environments, humanity and other animals.* Cullompton: Willan Publishing.

---

[24] As the author learned when he travelled in Indonesian Borneo near the border with East Malaysia (April 2005).

Bodard, K., & Pallemaerts, M. (2005). *Restricting the import in the EU of timber and timber products harvested through illegal logging: an exploratory legal review of available policy options*. Brussels: Institute for European Studies, Vrije Universiteit Brussels.

Boekhout van Solinge, T. (2002). *Drugs and decision-making in the European Union*. Amsterdam: Mets en Schilt.

Callister, D. J. (1992). *Illegal timber trade: Asia-Pacific*. Cambridge: Traffic International.

Carrabine, E., Iganski, P., Lee, M., Plummer, K., & South, N. (2004). *The greening of criminology. Criminology. A sociological introduction*. London and New York: Routledge.

EIA, & Telapak. (2004). *Profiting from plunder: How Malaysia smuggles endangered wood*. London: EIA.

EIA, & Telapak. (2005). *The last frontier. Illegal logging in Papua and China's massive timber theft*. London: EIA.

EIA, & Telapak. (2006). *Behind the Veneer: How Indonesia's last rainforests are being felled for flooring*. London: EIA.

European Commission. (2004). *FLEGT briefing notes*. Brussels: European Commission.

Frank, A. (2001). "Bush meat" crisis needs urgent action, group warns, National Geographic News, 22 May 2001.

Friends of the Earth. (2001). *European league table of imports of illegal tropical timber*. London: Friends of the Earth.

Global Witness. (2002). *The logs of war: The timber trade and armed conflict*. London: Global Witness.

Gray, M. A. (1996). The international crime of ecocide, *California Western International Law Journal, 26*, 215–271. Also reprinted in Nikos Passas (Ed.) (2003). International Crimes (pp. 455–511). Aldershot (UK): Dartmouth Publishing/Burlington (USA): Ashgate Publishing.

Greenpeace. (2001). *Logs of war. The relationship between the timber sector, and the destruction of the forests of Liberia*. Madrid: Greenpeace.

Greenpeace. (2005). *Danzer Group involved in bribery, illegal logging, dealings with blacklisted arms trafficker and suspected of forgery*. Amsterdam: Greenpeace Forest Crime File.

Hooghiemstra, D. (2002). In Gesprek over de nieuwe rijken met Jort Kelder, hoofdredacteur van Quote, NRC Handelsblad, 5 October 2002.

Jaakko Pöyry Consulting. (2005). *Overview of illegal logging*, Report prepared for the Australian Department of Agriculture, Fisheries, and Forestry. Melbourne: Jaakko Pöyry Consulting.

Leakey, R. E., & Lewin, R. (1996). *The sixth extinction. Patterns of life and the future of humankind*. London: Weidenfeld & Nicolson.

Mass Extinction Underway, Majority of Biologists Say (21 April 1998). Washington Post.

Naylor, R. T. (2005). The underworld of ivory. *Crime, Law and Social Change, 42*, 261–295.

Peal, A. (2000). Green spot in Africa. In K. Topfer (Ed.), *Our planet. The environment millenium*. UNEP.

Traynor, I. (2001). The international dealers in death, The Guardian, 9 July 2001.

UN Security Council, UN's Expert Panel Report on Sierra Leone, 20 December 2000 S/2000/1195.

Wilson, E. O. (2002). *The future of life*. New York: Vintage Books (Random House).

Ziel, A. van der. (2003). 'Tropisch hout met bloed eraan', in Dutch daily Trouw, 3 September 2003.

Chapter 9
The Role of Hawala Bankers in the Transfer of Proceeds from Organised Crime

Henk van de Bunt

Introduction

Hawala bankers[1] are financial service providers who carry out financial transactions without a license and therefore without government control. They accept cash, cheques or other valuable goods (diamonds, gold) at one location and pay a corresponding sum in cash or other remuneration at another location. This service is comparable with services provided by official banks and by non-banking financial institutions such as Western Union and MoneyGram, yet unlike these companies, hawala bankers disregard the legal obligations concerning the identification of clients, record keeping, and the disclosure of unusual transactions, to which these official financial institutions are subject.

Hawala banking has a long history. Primarily entrenched in the monetary facilitation of trade between distant regions, these bankers still provide a useful service especially to migrants who wish to transfer money to their country of origin. The informal system has long existed but only recently gained prominence in conflict-torn regions, such as Afghanistan (Maimbo, 2005) and Somalia (Houssein, 2005). After years of conflict, confidence in the formal banking system is absent and the remaining banks neither accept deposits nor extend loans. Significantly, these formal banks do not have the capacity to provide international or domestic remittance services (El-Qorchi, Maimbo, & Wilson, 2003: 9).

Despite the growing competition by formal remittance services, the use of hawala banking has probably not declined. According to a recent estimate by the IMF, migrants transfer $100 billion per annum to family members and relations in their country of origin through the official financial system. In addition, a similar amount of money is transferred in the form of goods, cash, and through underground bankers (International Monetary Fund, 2005). The IMF did not hazard an accurate

---

[1] Several traditional terms, like Hundi (India) and Fei-ch'ien (China) remind us of the fact that hawala banking sprang up independently in different parts of the world. At present, a range of other terms is used to refer to the same phenomenon, such as "informal banking," "underground banking," "ethnic banking" or "informal value transfer system."

D. Siegel and H. Nelen (eds.), *Organized Crime: Culture, Markets and Policies.*
© Springer 2008

estimate of the amount transferred through these informal banks. Presumably, the amounts are significant since hawala bankers are attractive to many migrants because they work more cost-effective, efficient, and reliable when it comes to transferring money (e.g. Perkel, 2004: 198 et seq.). An additional benefit for illegal migrants is that hawala bankers usually do not record their identity.

The recent tightening of control on formal financial channels has increased the vulnerability of informal banking to criminal abuse. The anti-money laundering effort has been a cornerstone of the fight against organised crime since the late 1980s. The objective is to prevent the misuse of the financial system by perpetrators of serious crimes. Anti-money laundering legislation has made the use of official financial channels less safe for offenders. From their perspective, transferring money through banks or non-banking financial institutions leaves hazardous paper trails and entails risks.

Since 9/11 the financing of terrorism has taken on a new urgency and has revitalised the efforts put into controlling the integrity of the financial sector. Although no direct links were found between hawala banking and the events of 9/11, there are indications that hawala banking has been used for the financing of terrorist attacks. The 9/11 Commission Report states that Al Qaeda frequently moved the money it raised via hawala (The National Commission on Terrorist Attacks, 2004: 171). However, the funds for the 9/11 attacks were not transferred through hawala banking, but either through wire transfers or in the form of cash (id: 172). The characteristics of hawala banking, including anonymity and lack of official scrutiny, make them susceptible to misuse (Johnston, 2005; Razavy, 2005: 288). Despite these concerns, not much empirical material on the misuse of hawala banking has been published to date.[2]

In this contribution, I will examine the use of hawala banking by criminal organisations, using the framework of a long-running research project on organised crime, the so-called Organised Crime Monitor (OCM). Since the start of this project (1996) up to now, 120 large-scale police investigations into organised crime have been analyzed. An earlier Dutch-language publication about this research (Kleemans, Brienen, & Bunt, 2002) provided an analysis of seven cases taken from police files of large-scale criminal investigations into informal banking.[3] Recently, a report was published based on 40 recent cases (Bunt & Kleemans, 2007). Three cases were primarily concerned with the transfer of criminal money by hawala bankers. In two other cases the connection with informal bankers was indirect, as the focus of the investigation was on the international drugs trafficking. In this contribution, I will mainly use the data from the three OCM-cases to illustrate the use of hawala banking for criminal purposes.

---

[2] Exceptions are, among others, Bunt & Dijken (2004), FATF (2003, 2005), Jost & Sandhu (2000), Kleemans et al. (2002), Passas (1999, 2005), and Thompson (2006).

[3] These cases are also used by Passas (2005).

Hawala Banking

Hawala banking is surrounded by myths and misunderstandings. For instance, it is claimed that hawala works by transferring money without actually moving it (Jost & Sandhu, 2000: 2).[4] This may be true in itself, but it is not characteristic of hawala banking only. With other kinds of international payments, including those that go through formal banking channels, funds are also remitted without being physically moved elsewhere (El-Qorchi et al., 2003: 14). It is also not a distinguishing feature of hawala banks that they transfer funds or value from place to place either without leaving a formal paper-trail of the entire transaction or without going through regulated financial institutions at all. According to Passas (2005: 17), it is a stubborn myth that hawala-type transactions are paperless and leave no trails[5]

The essence of hawala banking is simply that financial transactions occur in the absence of, or parallel to, formal banking sector channels. Hawala banking is not certified and supervised by any government. There are international differences in regulation and supervision of financial transactions by Central Banks (International Monetary Fund, 2005), but hawala banking has been legally banned in almost all countries of the world, even those which are strongly Islamic, such as Pakistan and Iran (Schramm & Taube, 2003: 409).[6]

To sum up briefly, informal bankers work outside the financial channels that are supervised and monitored by the competent authorities. The adjective "informal" is the most fitting to characterize this central issue. The informal operations of hawala bankers have far-reaching implications. First of all, hawaladers, unencumbered by formal rules, are able to provide a fast, efficient, and low-threshold service to their customers. No forms are filled out and no transaction needs to be recorded. For this reason, hawala bankers have a much lower overhead than formal institutions. According to the sociologist Razavy (2005: 279), this is the essential feature of the hawala system. But there is a second far-reaching implication. Because these financial transactions are not monitored and certified by the authorities (Central Banks), they rely heavily on the trust between clients and hawaladers, which brings us to the actual procedures and practices of hawala banking.

Hawala banking is a system in which money (or value) of the customer (*remitter*) is received by hawala banker *A* for the purpose of paying it to a third party (*recipient*) in another geographical location. *A* then communicates with hawala banker *B* at

---

[4] As Al-Hamiz (2005: 30) puts it, "Typically, a hawala transaction transfers the value of money from one country to another without the corresponding movement of cash or cover across borders."

[5] Curiously, at the same time, Passas (2005: 4) includes this lack of paper trails in his definition of informal banking.

[6] As far as the legal system in the Netherlands is concerned, informal bankers are engaged in activities contravening the Money Transaction Offices Act (*Wet inzake de Geldtransactiekantoren*), which prohibits the offering of money transaction services (currency exchange and money transfers) without a permit issued by the Central Bank (*De Nederlandsche Bank*).

the transfer destination requesting that funds be paid out to an individual identified by the initial customer. The communication may occur for example by telephone, facsimile or internet. The initial hawalader A charges the customer a fee or percentage of the transfer amount (circa 5%).

Trust is the most critical element in the system of informal banking. Without mutual trust between operators and clients, hawala banking cannot function. After all, hawala banker A has received funds in trust without making a payment, and B has made a payment without receiving its counter-value. Meanwhile, the remitter has to take it for granted that the money handed over to A will actually be paid to the recipient. In other words, trust between the two hawala bankers and between the hawaladers and their clients is essential.

Trust Between Hawala Bankers

Banker A receives money paid out by hawalader B, which leaves. A indebted to B. Hawala banking rests on the principle that the roles will be reversed in the next transaction: B receives money from a remitter to be paid out by A to the recipient. Now B owes A. When the amounts involved in both transactions are the same, the debts cancel each other out and there is no need for a physical transfer of money. In practice, the volume of transfers may be higher in one direction than in another. The hawaladers consequently need to settle their accounts by using conventional banking systems, physical transfers, or through a series of trade transactions (Kleemans et al., 2002: 120; Munshani, 2005; Passas, 1999).[7]

In theory, conflicts and misunderstandings about settlements are likely to arise, yet in practice relations between hawaladers appear to run smoothly. The economists Schramm and Taube (2003: 417) point to the fact that hawala networks are similar to "homogeneous clubs": networks which create lock-in situations through which extremely risk-laden exchange relationships can be transformed into self-fulfilling contracts. This can be explained by the strong social ties (ethnic or family) which usually structure the network of hawala bankers. Razavy (2005: 285) adds to this that the strength of the hawala banking system depends on its embeddedness in the common culture and belief system of the community. In this way, opportunistic behaviour by operators is likely to be doubly punished. Not only does an unreliable hawala banker risk expulsion from the hawala network, but also excommunication from the community and loss of honour (Razavy, 2005: 286; Schramm & Taube, 2003: 416). Broken trust results in economic suicide for the hawala banker, and

---

[7] Multilateral financial settlements involving several hawala operators located in different countries are more complicated.

because of these negative consequences hawaladers can rely upon the certainty that no one in the network would dare to cheat.

Trust Between Hawala Bankers and Clients

One of the most important characteristics of hawala banking is that the business relationship between banker and customer is based on *personal* trust. While many modern social arrangements tend to be secular and anonymous, the hawala system critically preserves the cultural mark of immediate and personal interactions (Razavy, 2005: 285). Sometimes these relationships grow so close that families deal with the same hawalader and his ancestors throughout generations; the constant encounter, together with the reputation that a hawala banker holds within a community, greatly delimits the likelihood of cheating. Schramm and Taube (2003: 416) also note that hawala bankers are strongly interwoven with the social-religious fabric of society. As a consequence of this particular embeddedness the norms and values affiliated with the entire system may gain a self-reinforcing element.

The criminologist Munshani, who interviewed 30 hawala bankers and 16 clients in the Afghan, Pakistani and Indian communities in Toronto, noted ethnic differences in the way "trust" plays a role in informal banking. In the Afghan community, it is trust with a strong feeling of kinship and honour towards the community. The fact that the hawala banker was an Afghan was enough for his clients to trust him and the banker in return wanted to help the Afghan clients. In the Indian community, on the other hand, religion and honour did not seem to come into play: "transactions were executed purely for economic gain" (2005: 10). In the Pakistani community trust was re-enforced with religion and the relationships had religious undertones (ibid.).

According to Munshani (2005: 13), a request of a receipt is offensive because it would doubt the credibility of the informal banker. Clients interviewed at these locations seemed uninterested in a receipt even if one was available. Considering the lack of formal procedures, the trust placed by clients in informal bankers is remarkably high. When, for instance, Surinamese immigrants in the Netherlands were asked why they use hawala bankers instead of official banks or money transfer agencies such as MoneyGram, they mentioned the trustworthiness of hawala bankers as an important consideration (Unger & Siegel, 2005: 91). In hawala banking clients and operators are often of the same ethnic background (Munshani, 2005; Perkel, 2004). Turkish migrants transfer money through Turkish informal bankers, while Surinamese migrants use Surinamese offices. This common background is in itself logical. First of all, sharing a common language and culture creates a basis of trust and, secondly, Turkish bankers are obviously in a better position to conduct money transactions in Turkey than, for instance, Surinamese or Iraqi bankers.

In conclusion, the intriguing aspect of hawala banking is that, as far as its customers are concerned, a maximum of trust is engendered despite a minimum of transparency and formality.

Use of Hawala Banking for Criminal Purposes

Besides the use of hawala bankers by migrants there is also the use by criminals. Hawala banking systems can be used for illegal purposes such as tax evasion; making or receiving payments in connection to crimes (kidnapping for ransom, human smuggling, drug trading, terrorism, etc.); and laundering the proceeds of crime. There are no reliable estimates of the extent to which hawala bankers, knowingly or unknowingly, assist in the misuse of their financial services. Judging by the number of publications, the interest in hawala banking increased considerably after 9/11 but, remarkably, misuse of the system usually receives little attention. Only in recent years have studies appeared into the use of hawala banking for criminal purposes.[8] A case in point is the development in the publications of the American criminologist Passas, who has written extensively on hawala banking. In earlier publications, Passas minimized the misuse of hawala banking. In 1999, he suggested that the growth of hawala banking for hard-core criminal purposes was limited by the wide availability of numerous other alternatives and the incapacity of informal networks to handle large sums of money on a regular basis. He even concluded that there was little evidence that the most sophisticated and organised criminals made use of informal banking systems (Passas, 1999: 68). However, in a recent publication, based on interviews and actual cases, he indicates that informal bankers are indeed capable of processing large sums of money for criminal purposes (Passas, 2005).[9]

Thompson's (2006) study on hawaladers in Afghanistan is one of the few examples of empirical research into misuse of informal banking. Thompson (2006: 185) interviewed 54 hawaladers in Afghanistan and she was able to show that fluctuations in the turnover of a large number of hawaladers corresponded to crucial phases in the cultivation and harvesting of opium. The obvious conclusion from this is that much of the money transferred through hawaladers is drug related.

In 2002, the Dutch Organised Crime Monitor published a report containing a.o. an analysis of seven cases of hawala banking (Kleemans et al., 2002). These cases demonstrated that hawala bankers regularly exchanged or transferred substantial amounts of criminal money. It also emerged that ethnicity was less important in

---

[8] See footnote 3.

[9] Case studies mentioned in a recent FATF report also demonstrate the capability of some groups of informal bankers to transfer large sums of money (FATF, 2003).

cases of misuse of informal banking for criminal purposes. The criminal clients of hawala bankers in these seven cases were usually of diverse ethnic backgrounds (id: 115–124). As part of a follow-up study by the Organised Crime Monitor, I examined the files pertaining to three large-scale police investigations into hawala bankers. I will first present these three cases in more detail and then offer some observations in section 4.

Case 1 involved a Pakistani informal banker working from a small shop in Amsterdam, who changed money and performed remittances almost exclusively for drug dealers of various nationalities (Afghan, Colombian, Dutch, and Venezuelan). He had been in the Netherlands for 17 years and had acquired Dutch citizenship. During the investigation his phones were tapped for a period of 7 months and detailed records of his financial transactions were found in a search of the premises. For every transaction (payments and remittances), he received instructions by telephone from an unknown person staying alternatively in Karachi and Dubai. In Amsterdam, the banker worked together with several other Pakistani who assisted him in his daily duties.

Case 2 also involved an informal banker in Amsterdam. This was a person of Indian descent who had been in the Netherlands for almost 20 years. He was engaged in currency exchange and money transfers, acting on behalf of both perpetrators of crime (drug dealers) as well as legal enterprises (clothing firms, car dealers) keen on dispatching "black money" out of sight of the tax authorities. He worked from a small takeaway restaurant, where more than 60,000 phone calls were tapped. During a search no company records were found and police interrogations established that no records of financial transactions were kept. In Amsterdam, the Indian banker collaborated with a number of people from his home region in India, but it was a Dutchman who was holding the "stash." The investigation shed little light on the nature and size of the international network within which the banker operated, apart from the fact that he did a great deal of business with a hawala banker in London.

Case 3 involved an Iraqi national engaged in numerous money transfers to Iraq and other international destinations on behalf of Iraqis in the Netherlands. The amount transferred per client was usually small. This informal banker was part of an international network of persons related by family (spanning Iraq, Germany, Sweden, the UK, Turkey, and Denmark). There was a lot of cooperation and personal contact between the various bankers in the network. In the course of the investigation approximately 17,000 phone calls were tapped. (Apparently the banker received or made a striking 200 calls per day.) During a search of his premises, records of financial transactions were found which revealed that around 90% of his activities concerned legal transfers by Iraqi migrants. The remaining 10% of the transactions related to human smuggling (organised by Iraqi nationals) and involved payments by smuggled persons to the smuggler.

In all three cases, the approximate turn-over of the hawala bankers was established. *Case 1* can serve as an example of the capability of hawala bankers to deal with large sums of money. The case is interesting because of its detailed reconstruction of the financial transactions conducted by a hawala banker during a period of

7 months. Police interrogations, tapped telephone conversations, and confiscated documents made it possible to record 147 financial transactions. These transactions involved international remittances, commissioned each and every time by the unknown key figure of the network (NN), who appeared to be constantly moving between Karachi and Dubai. The scheme amounted to NN ordering the Amsterdam informal banking office to make certain payments or receive a certain amount of money. For instance, a Columbian dealer would deliver English pounds, gained from the cocaine trade, to an informal banker in London, who would then inform NN of this delivery. NN would subsequently order the Amsterdam banker to pay the same amount in euros (minus commission) to another Colombian. A detailed reconstruction revealed that this particular informal banking office in Amsterdam received a total amount of about 6,636,000 euros and paid a total amount of approximately 3,270,000 euros. A commission of 6–8% was charged for every transaction. Settlement of accounts between Karachi/Dubai and Amsterdam was effected through physical transfers by couriers.[10]

In *case 2* no records were found, which made it difficult to establish the amounts that went through this informal banker's office. However, the earnings of one Antillean cocaine dealer, operating in the Netherlands and the UK, were assessed on the basis of currency exchanges and estimated at half a million euros in 6 months.

In *case 3* the investigation revealed that the informal banker transferred 1,588,000 euros over a period of 16 months and received 212,000 euros from abroad (mainly Iraq). The majority of these payments regarded legal transfers by migrants, but there were also payments by smuggled persons to human smugglers.

Discussion

More than Just Transferring Money

Besides informal banking systems there are other ways for the perpetrators of crime to transfer money abroad, out of sight of the authorities. The simplest method is the physical transportation of money with the help of couriers. Other possibilities are hiding money in cargo, converting money into items that can be moved easily such as diamonds, or using forged documents to take advantage of formal financial channels. Little is known about the extent to which these activities are taking place. There are indications that regular money transfers were widely used for a number of years, but that their popularity is dwindling, at least in the Netherlands, because of the cooperation by MoneyGram and Western Union with anti-laundering measures. Very recently, limitations were imposed to the option of carrying cash across the Dutch borders. But apart from these restrictive measures regarding alternative possibilities, hawala bankers have one great advantage: they do more than just transfer money.

---

[10] These sums were paid and received in different currencies, mainly pounds sterling, euros and rupees.

Not only do they provide financial services, but they also contribute to the smooth and safe settlement of criminal transactions.

In *case 2* an informal banker worked for a drug dealer (A) in the Netherlands, who sold drugs in the UK. One of his people (A1) received payment in the UK in sterling. Dealer A requested the hawala banker to make arrangements so that A1 could deliver his British pounds somewhere in the UK, and that he (A) could collect the amount in euros at the banker's small restaurant in Amsterdam.

The hawala banker later told the police that he then contacted an "acquaintance" (B) in the UK to make arrangements for A1 to deliver his British pounds. A1 received instructions from him to deliver the pounds at a certain place, to a man with a codename (B). In the event of A1 or B being questioned by the police, neither of them would be able to reveal the identity of the other. The only person holding the "key" would be the hawala banker in Amsterdam and as a result of these actions the "trail" of the money transfer was interrupted. Apart from hiding the trail, the banker also offered security to drug dealer A, because A would not have to worry about the financial goings-on in the UK and could confine himself to collecting the money in Amsterdam, minus a 6% commission for the banker.

Sometimes the role of hawala bankers is taken a step further. In *case 3* the banker played the same role as a notary in a real estate transaction in the legal world. The Iraqi banker was a man of stature within the Iraqi community and he only worked for Iraqi clients. His services to criminal clients were limited to Iraqi perpetrators of human smuggling. The banker oversaw the honest settlement of the illegal financial transactions related to human smuggling. The smuggler and his customers would agree on depositing the price of the transport with the hawala banker. After a successful smuggling operation, the customers or their families would contact the hawalader who would then hand the money to the smuggler.

To sum up, hawala bankers, either knowingly or unknowingly, play a larger part in illegal transactions than just providing a financial service. Apart from the advantages to criminals often mentioned in the literature (no identification of clients, no record keeping, etc.), there is another reason for criminal clients to use informal bankers: using hawala bankers can smooth out potential difficulties in the settlement of illegal transactions.

The Strength of Weak Ties

It is often cited in the literature on hawala banking that contacts between bankers and customers are based on trust; strong social ties seem to be the perfect answer to problems of trust or dormant mistrust between partners. The fact that family relations play an important role in networks of informal bankers and that clients often use bankers from the same ethnic background who live in the same neighbourhood, seems to confirm this. But strong ties in themselves are not entirely sufficient to avoid problems of trust. What matters in the end, is that a hawala banker can deliver on his promises. Strong ties can sometimes be an obstacle in this respect.

It is of great importance to hawala bankers that they earn a reputation for themselves. Reputation entails the public recognition of one's ability to make credible promises (Misztal, 1996: 121). When a good reputation has been established, the banker needs to act in accordance with his image as much as possible. A hawala banker with a reputation of being able to transfer large amounts of money within 2 days will do his utmost to live up to his name every time. He can not afford doubts about his capability. In this way, reputations can become straightjackets and paradoxical situations may occur.

In order to keep up one's reputation it is sometimes advisable to take chances, that is: to enter into financial transactions with persons outside the framework of strong social ties. According to Granovetter (1973, 1985), "weak" relations can be powerful. Weak relations are connections between people who are not intimate or close. But precisely because of this, these ties are not redundant and can have an important added value. Entering into new relationships may be risky in the context of illegal transactions, yet at the same time it can increase the effectiveness of the collaboration.

In *case 2* the hawala banker was approached by a major criminal client, who wanted a large sum of money changed from British pounds into euros. At the intercession of an acquaintance, the banker approached a Pakistani living in the UK, known to him only from phone conversations. The banker explained that he was mainly relying on the Pakistani's reputation of being able to change large amounts of money. In other words, business was conducted not primarily on the basis of trust in the Pakistani's integrity, but on trust in his (alleged) competence.

This observation is confirmed by Thompson's interviews with hawaladers in Afghanistan. She too concludes that the importance of strong social ties should not be exaggerated: "despite the emphasis on familial connections, it should be noted that the business is like any other, driven first and foremost by profit" (Thompson, 2006: 164). She also notes: "In reality, if it maximizes profit to deal with a Tajik when the dealer is Pashtun, he will not hesitate to do so after verifying that the other party is trustworthy. This is particularly the case in the trafficking of drugs" (id.).[11]

So, it appears that establishing client-banker contacts gets easier as the stakes get higher. As stakes get higher, shared ethnicity and a common social background play less of a role than is often assumed.

The Presence of Trails

The existing literature offers contradicting answers to the question as to whether or not hawala bankers leave paper trails. According to some authors, informal bankers typically leave no traces of financial transactions behind, either because nothing is

---

[11] See also Passas (2005: 23). He concludes that in a globalizing economy and in view of very substantial amounts of money going around the globe, informal bankers must rely on people beyond kinship or even ethnic ties.

recorded, or because records are destroyed immediately after the completion of a transaction (Jost & Sandhu, 2000: 3; Perkel, 2004: 202; Razavy, 2005: 279). Others claim that hawala bankers document their transactions or keep records (Kleemans et al., 2002; Passas, 2005: 17; Thompson, 2006: 165). Regardless of whether paper trails are left or not, there will always be trails. Ironically, the lack of record keeping leaves traces, precisely because the absence of written confirmation is often compensated in practice by intensive telephone traffic. Informal bankers are constantly on the phone, discussing who gets paid when, where, and at which exchange rate (transfers are often paid out in another currency).[12] Hawala bankers cannot afford to make mistakes as a result of unclear arrangements.

The reputation of hawala bankers is crucial to their success. Reputations are created through talk and gossip; they are sustained and modified by face-to-face interactions. To build up a reputation as a reliable hawala banker within the community, hawaladers have to operate punctually and predictably: they will use fixed mobile telephone numbers, and visit regular places at regular hours. An example of "reputation-management" can be found in a generous gesture by a Pakistani banker (*case 1*). Together with a client, an Afghan hashish dealer, he was on his way to a colleague to change 85,000 pounds into euros, when he was robbed and almost killed by bullets. Tapped telephone conversations revealed that the banker played down his loss to the outside world on the one hand (maintaining he was only robbed of 6,000 pounds), while on the other hand he felt honour bound to compensate for the Afghan's loss.[13]

Conclusions

There are at least two different perspectives on hawala banking. From one point of view, hawala banking is regarded as a centuries-old institution which has not yet outlived its usefulness. Low-income workers and migrant workers in particular supposedly put more trust in hawala bankers than in formal banks. This viewpoint emphasizes the problems associated with subjecting hawala banking to the same rules as formal banks. Regulation either through registration or licensing is seen as ineffective because it will simply push the system farther underground, further complicating the already problematical task of controlling hawala transactions (Perkel, 2004: 210–211; Razavy, 2005: 292).

From the opposite perspective, hawala banking is described as "underground banking," a system that flies under the radar of modern supervision of financial

---

[12] The exchange rate is also part of the profit hawala bankers make. Currency trading and negotiating exchange rates is part of their business.

[13] Passas (2005: 19) mentions the example of the reputation of a network being at stake. An informal banker lost a fortune in gambling and went bankrupt. A meeting took place between the informal bankers involved and arrangements were made to cover and support each other. No retail customer lost a penny. The bankrupt partner was ostracized and ended up as a taxi driver.

transactions. Underground banking is considered a threat to the effectiveness of anti-money laundering measures and the fight against terrorist financing. To prevent underground bankers from becoming a safe haven for criminals and terrorists, they should be subject to the standard regulations regarding record keeping, disclosure of unusual transactions and identification of clients.

Because of these conflicting perspectives, the debate on the usefulness and necessity of hawala bankers runs the risk of being taken over by ideology. This is why it is important to carefully analyse hawala banking before suspicions and clichés are created. Up to now, little empirical research has been carried out into the modus operandi of hawala bankers and their views on their own profession. For instance, it is not clear to what extent they screen potential clients themselves, in order to prevent misuse of their services. The proportion of transfers for criminal purposes to legitimate transfers by migrant workers is also unknown. Is misuse only a marginal phenomenon when compared to the total number of legitimate transactions?

In this article, I have examined the role of hawala bankers starting from the research of the Dutch organised crime monitor. Of course, the above-mentioned cases tell us nothing about the pervasiveness of the misuse of informal banking, but they do reveal something about the nature of the cooperation between hawala banking and criminal organisations. The cases challenge a number of existing preconceptions and clichés about hawala banking.

It turns out that hawala bankers do leave paper trails and other trails; the idea that informal transactions leave no trace behind is largely untrue. Underground bankers are also capable of transferring large sums of money at short notice. Furthermore, the service provided by hawala bankers is not limited to transferring money. It appears from these cases that hawala bankers, either knowingly or unknowingly, play a larger part in illegal transactions than just providing a financial service. Hawala bankers can smooth out potential difficulties in the settlement of illegal transactions (witness the example of the human smuggler and his customers who agreed on depositing the price of the transport with the hawala banker). Finally, the importance of ethnicity and personal trust should not be exaggerated. In several cases, the bankers and their (criminal) customers were of different ethnic origins and contact was established through other than ethnic channels. When the stakes are high, a common social background and shared ethnicity between different bankers, as well as between bankers and their clients, seem to be less important than is often assumed.

Acknowledgment The author would like to thank Dina Siegel for her useful comments.

References

Al-Hamiz, S. (2005). Hawala: A U.A.E. perspective. In International Monetary Fund, Monetary and Financial Systems Department. *Regulatory frameworks for hawala and other remittance systems* (pp. 30–35). Washington, DC: International Monetary Fund.

Bunt, H. G. van de, & van Dijken, A. (2004). Official and informal financial services. In H. G. van de Bunt & C. R. A. van der Schoot (Eds.), *Prevention of organised crime: A situational approach* (pp. 55–62). Den Haag: Boom Juridische Uitgevers.

Bunt, H. G. van de, & Kleemans, E. R. (2007). *Georganiseerde criminaliteit in Nederland. Derde rapportage op basis van de monitor georganiseerde criminaliteit.* Den Haag: Boom Juridische Uitgevers.

El-Qorchi, M., Maimbo, S. M., & Wilson, J. (2003). *Informal funds transfer systems: An analysis of the informal hawala system.* Washington, DC: International Monetary Fund, Occasional Paper No. 222.

FATF. (2003). *Combating the abuse of alternative remittance systems. International best practices.* Financial Action Task Force on Money Laundering.

FATF. (2005). *Money laundering & terrorist financing typologies 2004–2005.* Financial Action Task Force on Money Laundering.

Granovetter, M. (1973). The strength of weak ties. *American Journal of Sociology, 78*(6), 1360–1380.

Granovetter, M. (1985). Economic action and social structure: The problem of embeddedness. *The American Journal of Sociology, 91*(3), 481–510.

Houssein, M. D. (2005). Somalia: The experience of hawala receiving countries. In International Monetary Fund, Monetary and Financial Systems Department. *Regulatory frameworks for hawala and other remittance systems* (pp. 87–93). Washington, DC: International Monetary Fund.

International Monetary Fund. (2005). Globalization and external imbalances. *World economic outlook: Globalization and external imbalances* (Chap. 3, pp. 69–84). Washington DC: International Monetary Fund.

Johnston, R. B. (2005). Work of the IMF in informal funds transfer systems. In International Monetary Fund, Monetary and Financial Systems Department. *Regulatory frameworks for hawala and other remittance systems* (pp. 1–6). Washington, DC: International Monetary Fund.

Jost, P. M., & Sandhu, H. S. (2000). *The hawala alternative remittance system and its role in money laundering.* Lyon: Interpol General Secretariat.

Kleemans, E. R., Brienen, M. E. I., & van de Bunt, H. G. (2002). *Georganiseerde criminaliteit in Nederland. Tweede rapportage op basis van de WODC-monitor.* Reeks Onderzoek en Beleid 198. Den Haag: WODC/Boom Juridische Uitgevers.

Maimbo, S. M. (2005). Challenges of regulating and supervising the hawaladers of Kabul. In: International Monetary Fund, Monetary and Financial Systems Department. *Regulatory frameworks for hawala and other remittance systems* (pp. 47–64). Washington, DC: International Monetary Fund.

Misztal, B. A. (1996). *Trust in modern societies.* Cambridge: The Polity Press.

Munshani, K. (2005). *The impact of global international informal banking on Canada.* An Empirical Research Commissioned by The Nathanson Centre for the study of organized crime and corruption and the Law Commission of Canada.

The National Commission on Terrorist Attacks. (2004). *The 9/11 Commission report. Final Report of the National Commission on Terrorist Attacks Upon the United States.* New York: W. W. Norton & Company.

Passas, N. (1999). *Informal value transfer systems and criminal organizations: a study into so-called underground banking networks.* Den Haag: Ministerie van Justitie, WODC.

Passas, N. (2005). *Informal value transfer systems and criminal activities.* Den Haag: Ministerie van Justitie, WODC.

Perkel, W. (2004). Money laundering and terrorism: informal value transfer systems. *American Criminal Law Review, 41*(1), 183–211.

Razavy, M. (2005). Hawala: An underground haven for terrorists or social phenomenon? *Crime, Law and Social Change, 44*(3), 277–299.

Schramm, M., & Taube, M. (2003). Evolution and institutional foundation of the hawala financial system. *International Review of Financial Analysis, 12*(4), 405–420.

Thompson, E. (2006). The nexus of drug trafficking and hawala in Afghanistan (Chap. 6). In D. Buddenberg, & W. A. Byrd (Eds.), *Afghanistan's drug industry: structure, functioning,* dynamics and implications for counter-narcotics policy (pp. 155–188). Washington, DC: United Nations Office on Drugs and Crime (UNODC) and The World Bank.

Unger, B., & Siegel, M. (2005). *The amounts and effects of money laundering.* Report for the Ministry of Finance, February 16, 2006. Utrecht school of Economics.

The World Bank. (2006). *Global economic perspectives. Economic implications of remittances and migration.* Washington, DC: The International Bank for Reconstruction and Development/ The World Bank (published November 2005).

Chapter 10
Facilitating Organized Crime: The Role of Lawyers and Notaries

Hans Nelen and Francien Lankhorst

On Monday, 31 October 2005, 36-old Dutch lawyer Evert Hingst was murdered in front of his house in Amsterdam. Among Hingst's clients were many noted and alleged criminals, including John Mieremet, who was once shot in front of Hingst's office. After this assault, Mieremet claimed that Hingst had tried to set him up. Mieremet was murdered 3 days after the liquidation of Hingst, on 2 November, in Thailand. Hingst, a fiscal specialist, had been accused of assisting criminals to launder their money abroad. He was imprisoned for several weeks after police discovered three firearms and a large sum of cash during a raid of his office in 2005. A suspect of money laundering, membership of a criminal organisation and possession of firearms, Hingst gave up his profession as a lawyer in July 2005. He had previously been arrested on charges of forgery of documents in 2004.

The case of Evert Hingst, although exceptional in many ways, reflects the fact that people who render financial and legal services may play a vital role within and between criminal networks. Recently, at both national and international level, the compromising conduct of these facilitators has been the subject of public discussion, as have the possible measures that should be taken against such conduct.

The discussion regarding the role of facilitators is best illustrated by the update and expansion of the Money Laundering Directive dating from 1991.[1] This bitterly contested 2001 Amending Directive,[2] approved by the European Parliament in the emotional aftermath of "9/11," stipulates that member states should compel several professional groups, including lawyers and notaries, to report suspicious financial transactions to the appropriate authorities. Notably, this has had far reaching implications for lawyers and notaries and their professional attributes, such as the legal exemption to testify against their clients about any information entrusted to them on a confidential basis.

But what do we actually know about the nature and extent of unacceptable links between professionals who render legal and financial services on the one hand, and

---

[1] Directive 91/308/EEC of the Council dated 10 June 1991 on prevention of the use of the financial system for the purpose of money laundering, *PbEG* 1991 L 166/77.

[2] This refers to Directive 2001/97/EC of the European Parliament and of the Council of 4 December 2001, amending Council Directive 91/308/EEC on prevention of the use of the financial system for the purpose of money laundering. This guideline was published in OJ L 344, 28 December 2001, 76.

D. Siegel and H. Nelen (eds.), *Organized Crime: Culture, Markets and Policies.*
© Springer 2008

criminals on the other hand? The classical, stereotype image is the *consiglieri* who is part of a criminal network and operates as the most important consultant of the head of a crime family. Although both nationally and internationally the number of studies into the facilitating role of lawyers and notaries is rather small, the studies that are available have demystified this fallacy. In the USA, Green (1990) showed that lawyèrs have abundant opportunities to commit crimes in the course of their occupation, but the cases he referred to were merely examples of violations of fiduciary duties and violations of codes of conduct, rather than straightforward examples of symbiotic relationships with organized crime. In the Netherlands, Van de Bunt (1996) was the first criminologist who systematically charted the links between professionals and organized crime. With regard to lawyers, he did not find any case that came close to the image of *consiglieri*. However, he concluded that, unlike accountants, tax advisers and notaries, culpable involvement in organized crime activities by lawyers was more than incidental. In his report, Van de Bunt described 29 cases of culpable involvement of lawyers and 12 cases of compromising conduct of notaries. Two years later the first report of the so-called "Dutch organised crime monitor" of the Research and Documentation Centre (WODC) confirmed the finding that some professionals facilitate wrongdoing by others, including organized criminal groups (Kleemans, van de Berg, van de Bunt, 1998).

The contribution in this chapter gives an overview of contemporary manifestations of compromising conduct of lawyers and notaries in relation to organized crime.[3] The cases described were collected and analysed during an international research project conducted in four European countries: Italy, France, the United Kingdom, and the Netherlands.[4] The definition of culpable involvement developed by van de Bunt (1996) was taken as a starting point. Hence, two forms of culpable involvement are distinguished. Firstly, a professional can be involved in punishable acts in such way that he himself can be criminally prosecuted. He can aid and abet the punishable acts of criminal networks, or facilitate them, for example by fencing/screening stolen goods, or laundering money. This is the strictest sense of culpable involvement.

Secondly, there is culpable involvement in a broader sense if the professional fails to exercise due care to prevent the misuse of his professional capacity for criminal purposes. This does not necessarily imply conscious and deliberate

---

[3] The United Kingdom does not have an independent profession of notaries. The function, the tasks and the competence of a number of solicitors and so-called 'licensed conveyancers' are partly similar to those of the notaries in the other countries.

[4] The research in Italy was conducted by Paola Zoffi, Andrea di Nicola and Ernesto U. Savona (on behalf of Transcrime, Università di Trento – Università Cattolica di Milano); in France by Emmanuelle Chevrier and Michel Massé (Université de Poitiers); in the United Kingdom by Michael Levi and David Middleton (Cardiff School of Social Sciences, Cardiff University); in the Netherlands by Francien Lankhorst and Hans Nelen (Faculty of Law, Department of Criminology, Vrije Universiteit Amsterdam). The final report of the research was submitted to the European Commission but was never published. However, in a special issue of *Crime, Law and Social Change*, 42(2–3), 2004, the most important findings of the study were presented.

collaboration resulting in criminal offences, but the reproach can be voiced that he could or should have known that his services were being wrongfully used for criminal purposes. It should be clear that the standard of due care goes further than what is punishable by criminal law.

The focus in this chapter is on the *nature* of the problem. It is very hard to draw any reliable conclusions with regard to the extent of culpable involvement of professionals. Although it has to be emphasised that the number of lawyers and notaries that can be demonstrated to be involved in wrongdoing is small in comparison to the size of the professions as a whole, for several reasons the findings of the international study in no way prompt trivialisation of the problem.

Firstly, it is likely that an unknown number of cases of culpable involvement is never investigated. Due to the special position of the legal and notarial profession in society, the law enforcement agencies in most countries exercise the utmost reluctance and caution in starting investigations against members of these professions.

Secondly, the lack of relevant financial expertise among law enforcement officials inevitably has an effect on the number of cases that have come to light. Indeed, where there is a lack of expertise and skills to unravel complex money laundering transactions, getting to the bottom of the possible dubious role of a lawyer or notary in this context, is also problematic.

Thirdly, it must be born in mind that financial and legal service providers do not usually work exclusively for a single criminal organisation, but make their services available to various criminal organisations (Chevrier, 2004; Kleemans, Brienen, & van de Bunt, 2002; Lankhorst & Nelen, 2004, Middleton & Levi, 2004; Nicola & Zoffi, 2004). Hence, the actions of a low number of lawyers and/or notaries who exceed the limits of the acceptable, have the potential to not only inflict (and indeed have inflicted) significant damage on the professions' reputation as a whole and their finances, but also to disrupt society.

Before sketching the various manifestations of culpable involvement by lawyers and notaries, a number of factors are taken into account that may influence professional standards within the relevant professional groups and may increase the risk for lawyers or notaries to become culpably involved in (serious) forms of crime.

Risk Factors

The combination of financial and legal knowledge, and the aura of respectability and reliability make both lawyers and notaries attractive potential partners for organized crime. After all, they are the ones who can give criminal entrepreneurs legal advice, and can provide the expertise needed to launder criminal proceeds. To the outside world, the participation of a lawyer and notary in a transaction can create the impression of legitimacy. From the perspective of the criminal network, the most important surplus value of the professional concerns the confidential nature of the relationship and the privilege of non-disclosure. This is the main reason why interactions between lawyers and notaries on the one hand, and their clients and associates on the other hand, are inherently non-transparent (Fijnaut, et al., 1998).

Some of the most significant changes that have taken place within the legal and notarial professions in France, Italy, the United Kingdom and the Netherlands over the past 30 years are the growth in number, prestige and power of the legal professions, the globalisation of the business and capital markets and the growth and ever-increasing sophistication of organized criminal groups that operate across national borders. Operating together, these changes could constitute a serious threat to the integrity and the ethics of the professionals and create distinctive cultural, ethical and legal challenges, especially for lawyers and notaries.

The process of the globalisation of trade, finance, communications and information has resulted in profound changes in the business and capital market. Businesses have begun to reach beyond national borders and offer their products and services to individual and government consumers in all parts of the global village. Worldwide marketing, selling, and purchasing has become a matter of financial survival. This shift added a new dimension (and burden) to the legal and notarial profession. In addition to representing clients in and out of court and giving legal advice, a large number of lawyers tend to engage more and more in other commercial activities. They may create corporate and trust entities (in mainstream jurisdictions and offshore), act as a manager of these entities, open bank accounts or conduct financial transactions on behalf of their clients etc. The distinction – in terms of commercial activities – between lawyers, notaries and other professionals, such as tax consultants and other financial and legal specialists, has faded somewhat during the last decades. The major difference that remains is the privilege of non-disclosure. This characteristic makes lawyers and notaries an attractive bridgehead for organized crime to enter the legitimate economy.

Internationalisation

Over the past several decades the commercial section of the profession of law has been marked with a previously unknown expansion frenzy and up-scaling. Initially this would lead to joint ventures and mergers on the domestic front but gradually branches have been opened abroad and have merged with foreign firms.

Regarding integrity, internationalisation has had both positive and negative effects on the legal profession. Positive effects derive from the fact that checks on integrity are further developed under the Anglo Saxon system than in France, Italy and the Netherlands. The USA and the UK have higher liability risks and this encourages attention to and discussion on themes touching on the integrity of the professional groups. Moreover, internationalisation has meant that larger firms usually deploy several lawyers on the case of a single client. This generates a more natural form of control among colleagues.

On the other hand, increased internationalisation and up-scaling can also impact negatively on transparency. Hence, it will not always be immediately clear which client an internationally operating firm is dealing with, what the client is up to, and the jurisdiction covering contacts between lawyer and client.

Commercialisation

Internationally, a trend can be observed that the number of practitioners who are entitled, under certain circumstances, to register themselves as lawyers, is increasing. In a sense, this development has made the difference between lawyers and other providers of business services increasingly diffuse. A split becomes apparent in the legal profession, whereby the "big money" is earned in the consultancy branch and pleading in court and legal advice for private individuals is in danger of sinking to the lower end of the market.

Due to growing competition and erosion of the market position, (bona fide) companies may switch to unethical practices to survive. Criminal networks are skilled in identifying these "weak links." Whilst it is beyond the scope of this study to assess trends concerning the financial viability of legal practice, it is likely that wrongdoing by lawyers and notaries is linked to economic issues, at both the micro and macro levels. In particular during an economic downturn the risk has to be taken into account that some firms become too dependent on relations with a limited number of large clients. It is not inconceivable that, in order to survive economically, the integrity of these firms may come under pressure.

Types of Manifestations of Culpable Involvement

The previous section, mainly hypothetical in character, sketched a number of risk factors whereby lawyers and notaries, in addition to other service providers, may be at a higher risk of involvement with given types of organized crime. Using research material from literature, interviews and dossier investigations, this section will describe the forms of culpable involvement actually encountered in the four countries in which the research was conducted.

Lawyers

Culpable involvement of lawyers is often associated with lawyers specialising in criminal defence law. This is because their clients, the type of case dealt with, and the related publicity put them more often in the spotlights than their civil-law colleagues. However, in contrast to the general idea, most cases of compromising conduct are related to lawyers trained in civil law. Firstly, the types of culpable conduct where mainly criminal defence lawyers are involved will be outlined, followed by some illustrations of cases dominated by civil lawyers.

"Mailbox" or "Errand Boy"

In various cases – in France, Italy and the Netherlands – the lawyer acts as a "messenger" or "mailbox" by passing information that should remain unknown to certain persons, to his client, the client's family or other parties.

> In this context, the Italian national report refers to one of the mafia-style maxi trials in Turin, in which the public prosecutor started an investigation and found evidence that a lawyer was acting as a go-between for several members of the organized criminal group in custody and other suspects and wanted accomplices.
> The Dutch report refers to a situation in which a criminal defence lawyer had several clients, simultaneously, in the same case. During the process whereby suspect A was being put in police custody, the lawyer passed information to suspect B, who had not yet been arrested at that time. Moreover, the lawyer also allowed suspect A, who was subject to restrictions, to read the statements of other suspects.[5]

According to members of the Dutch legal profession, in general a lawyer does not contravene due care if he acts as messenger for his client, and shuttles between members of a criminal organisation. These members of the profession claim that contacts between a client and the lawyer may extend to a considerable degree. This also includes contacts with other possible witnesses and suspects. Indeed, as a lawyer, one is required to serve the interests of the client, on a partisan basis. In practice none of the lawyers interviewed had encountered a case whereby a "boss" of a criminal group, having been detained, used the lawyer to pass on instructions to the "small-fry" as to what they should say. Similarly, they had no personal experience of passing on sealed envelopes. At the same time, they do not consider this practice as reprehensible under disciplinary regulations, by definition, although they recognise that it means an increased risk of acting as messenger for matters in which lawyers would prefer not to be involved (Lankhorst & Nelen, 2004).

The views of the lawyers are in sharp contrast with the perceptions of law enforcement officials. In their opinion, it is quite common for information to be passed to unauthorised persons. Relevant documents are released 3 days after an arrest. These often feature the names of persons who are still the focus of an investigation. On occasion, this information has been passed by the lawyer to an unauthorised third party (Lankhorst & Nelen, 2004).

There is another way in which information can be passed to unauthorised third parties. Occasionally, in the interest of the client, a lawyer will put (some of) the client's criminal file on the internet or show this to journalists. Particularly in a case

---

[5] Subjection to restrictions means being detained whereby all contact with the outside world is prohibited except via the lawyer.

with several accused, this can breach confidentiality and possibly also interests around legal proceedings

Intimidation or Manipulation of Witnesses

A second manifestation of culpable involvement is the situation where the lawyer pushes his partiality to extremes. The criminal defence lawyer tries to influence or to intimidate a witness in order to persuade him to withdraw incriminating statements against his client.

> The Italian report reveals a case in which the lawyer had asked a key witness to mitigate his accusatory testimony against his client who was accused of extortion.
>
> A Dutch case relates to an attempted crime of aiding/abetting. The lawyer had tried to exert pressure on a defendant, who was not his client, to persuade him not to mention some factual circumstances linking his client to possible wrongdoing.

Assistance to Human Trafficking

A relatively new phenomenon is the assistance that some lawyers provide in cases related to human trafficking. Human smuggling has attracted a great deal of official attention in the past decade and official documents stress the importance and responsibility of human smugglers for the nature and size of migration flows, especially the influx of asylum seekers in European countries.

The research material indicates that, in particular with regard to the supply of false identity papers, good contacts with members of the legal profession may be useful for human smugglers. In particular in Italy, cases have been found in which lawyers provided services with regard to the organisation of clandestine immigration.

> The Italian national report refers to the arrest and conviction of a lawyer accused of forgery, slander and breaking immigration laws. He had counterfeited documentation, allowing several non-EU citizens to obtain residence permits to which they had no rights. Through forced labour contracts, the lawyer attested the presence of the non-EU citizens in Italy before January 1998, a requirement necessary to benefit by the deed of indemnity. The lawyer was also convicted of the slander of several policemen, declaring with written statements and during an interrogation that they had abused their powers and slandered him and his clients.
>
> In February 2000, an Italian lawyer was convicted to 5 years and 6 months imprisonment for having organized clandestine immigration. In November of the same year, a criminal lawyer was arrested in Modena. The lawyer laundered the proceeds of trafficking in women. The trafficking was organized by a band of Albanians who were connected to local criminal organisations. They imported women from Albania and forced them into prostitution with threats and violence, in effect reducing them to slavery. The Albanian group, with the aid of the lawyer, reinvested the proceeds in economic activities, above all in Albania. The lawyer was accused of money laundering and external participation in a criminal association with the aim of forging public documents.

The Italian and Dutch case studies further suggest that legal advice on marriages of convenience may be very important for criminals involved in trafficking of women.

Finally, in the national report of the United Kingdom fraudulent arrangements for obtaining permission of asylum seekers to remain in the UK are described.

> A British solicitor was convicted of offences related to making fraudulent arrangements for obtaining permission of asylum seekers to remain in the UK. The case clearly falls within the EU definition of organized crime in that the solicitor was working with at least two others, one of whom was reported in the press as being an "enforcer" and "member of the notorious Snakehead gangs that smuggle Chinese citizens to Britain."

Abuse of Facilities of the Law Firm

Noteworthy are also the cases in which the lawyers gave criminals ample opportunity to abuse the facilities of the law firm – phone, fax, e-mail – to communicate with each other. From the point of view of criminals, this is a perfect instrument to shield criminal activities. After all, both criminals and lawyers are well aware of both the legal thresholds and the reluctance amongst police officers and magistrates to use special police methods against law firms. Due to the fact that, unless the lawyer is the main suspect of an investigation, (tele) communications from a lawyer's office will not be intercepted, criminals can freely exchange information.

> The Dutch national report refers to the situation where a lawyer allows his client A to make use of the different means of telecommunications at his office. When A is arrested, it appears that A, through the means of the lawyers' telecommunication, intensively contacted other suspects. Because of the possibility to make use of the means of telecommunication at the lawyers' office, crimes can be prepared and or executed, whilst being relatively well shielded.

Conflicting Representation

Partiality is the duty of the criminal defence lawyer. In case the lawyer represents several clients at the same time, conflicts of interest may occur. In all four countries, the codes of conduct contain a ban on conflicting representation. In general, if meaningful agreements are made with all parties and their attention is drawn to the possible consequences, a single lawyer can act for several parties with possibly conflicting interests. However, the case studies reveal that sometimes criminal defence lawyers officially act on behalf of two or more parties, while in fact they only fully represent the interests of one of the parties. As a consequence, they severely inflict the interests of the other clients.

A particular situation exists in Italy. Criminal defence lawyers in this country can play an important role in criminal proceedings regarding organized crime, corruption and company crimes (*reati societari*). In these types of proceedings, it can happen that the same lawyer represents several of the accused. A single defence lawyer means a single defensive line for all, which is oriented towards the protection of the association, more than to protecting the interests of a single individual.

If an individual opts for the group defence line, there are no problems. However, case studies show that difficulties arise when an individual wants to be dissociated from the criminal group.

> The Italian report in particular refers to the risks a criminal defence lawyer may run, once he decides to represent a *pentito* (super-grass). If a prisoner intends to collaborate with the judicial authorities, the magistrate has to appoint him a public defender. It turns out to be difficult, sometimes even impossible, to find a lawyer willing to engage in such work. Any lawyer willing to represent this type of client would find himself placed on a black list by organized crime, which could cause a loss of income, proportionate to the importance of the super-grass. What's worse, he could become subject to serious forms of intimidation.

Money Laundering

Concerning money laundering, the case studies predominantly reflect the involvement of lawyers with regard to giving advice on beneficial ways to transfer money, thinking up (fake) constructions, and passing (criminal) money through their accounts. In a number of cases, the lawyer was inserted into a chain of financial transactions and provided several of the aforementioned services. They not only stored large sums of money on behalf of their clients, but also passed money through their own personal or business accounts to other accounts, and thought of complex financial constructions to mislead the judicial and fiscal authorities.

In some of the cases, we can observe a shift in the role that the lawyer plays within the criminal network. He starts as an external expert who is functionally connected to the criminal network, but slowly becomes more and more involved, both commercially and socially. He joins the leading members on social events – parties, football matches etc. – and regularly and deliberately performs all kind of services for them. The case of Evert Hingst, described at the beginning of this chapter, is a good example of this development.

> Another illustration is a Dutch drugs case in which the two main important experts are a lawyer and a legal advisor. The first advises the leading members of the criminal group on a regular basis. He has his own licit legal practice, which is located in the office complex of one of the leading members of the network.

Third Party Account as Safe-Haven

Lawyers have third party accounts, used for the money flow between the different parties with the professional as intermediary. The funds in these accounts might be susceptible to improper use by the professionals. One can think of unauthorised payments to third parties, or using the third party account as a safe haven, because of the fact that it is difficult to seize funds placed in these accounts. Moreover, transferring money through the professional breaches the trace, which makes it difficult to get at the source of the money flow.

In the United Kingdom, the ability of solicitors to transfer funds anonymously through client accounts turns out to be a vulnerable spot.

Several cases in the United Kingdom were identified in which the proceeds of a mortgage fraud upon a lending institution were distributed (in effect, laundered) in accordance with the fraudsters' wishes. Since the money was transferred by a solicitor, most recipients assumed that the transaction was legitimate.

Involvement in Other Forms of Financial Economic Crime

The case studies and interviews confirm the notion that fraudsters sometimes try to involve professionals in their activities in order to give these activities credibility. The data from the United Kingdom in particular contain several examples of misconduct of legal professionals in relation to fraud schemes.

Mortgage Fraud[6]

Lawyers in the UK are *involved* in mortgage fraud primarily for the structural reason that it involves the conveyancing of land which, when carried out for a fee, is an activity reserved to their profession.[7] Lawyers become *implicated* in mortgage fraud because in the current most common form of conveyancing, they act for both the buyer (the borrower of funds to finance the purchase) and for the lender, and the fact that the solicitor has two clients creates potential for conflicts of interest. That is not, however, the whole story. A fraudster intent on obtaining a loan (secured by mortgage on land) by deceiving a lender could quite lawfully carry out his own conveyancing, but in such circumstances the lender would instruct solicitors to act on its behalf, would be careful to ensure that the loan was properly secured and would check that the details of the transaction were genuine. This introduces a key theme to lawyer-facilitation of wrongdoing: the advantage to the fraudster is that the involvement of a solicitor adds *credibility* to the transaction. In a sense, it causes the lender to drop its guard. The lender assumes that since it is also the lawyer's client (which is how most transactions are structured), he will protect its interests. The lender is also aware that the lawyer carries compulsory indemnity insurance[8] and, in the case of solicitors, that the Solicitors' Compensation Fund (SCF) is available as a last resort from which it can seek to recover its loss.[9]

---

[6] This section is primarily based upon the report and article of Middleton and Levi (2004).

[7] As will be described below, the same applies to notaries in Latin-orientated law systems.

[8] The Solicitors' Indemnity Insurance Rules (made annually), pursuant to section 37 of the Solicitors Act 1974.

[9] Section 36 of the Solicitors Act 1974.

Investment Fraud

Cases of investment fraud in the United Kingdom show that reassurance to victims that their investments are protected by lawyers' compulsory indemnity insurance and compensation funds is also useful in persuading them to release their money. This is often effected in practice by the investor's funds being transferred to a lawyer's client or trust account where it is supposed to remain until receipt of the fantastic returns promised. Over twenty solicitors have been struck off because of their involvement to some degree in these schemes, and many firms have been subject to intervention for the same reason. The next example illustrates how a lawyer may be involved.

> In this case, over US$12 million passed through the hands of the solicitor. He and his associates promised loans of up to US$50 million to borrowers in exchange for advance fees of, in one case, US$3.5 million. The advance fees were then stolen. The solicitor was sentenced to 9 years in prison, was bankrupted, and he was struck off. The borrowers had been concerned to "ensure" that the fee they paid for the loan was secure, so insisted that it be paid to a solicitor in England and Wales. This provided an illusion of security.

The sums of money passing through solicitors' firms in relation to these transactions can be very substantial. For example, in one case it was reported that in a period of 7 months over US$50 million had been received by one firm and the solicitor himself told the investigator that "US$16,000,000 had apparently been misappropriated."

Notaries

Four forms of culpable involvement were identified in the case studies and interviews with regard to notaries. All these manifestations were also highlighted in recent reports of the Financial Action Task Force (FATF) on Money Laundering Typologies. The cases of the notaries are related to either transactions on the property market, or the creation of corporate vehicles and other complex legal arrangements. No cases of culpable involvement of notaries – that is: in relation to organized crime – were found in the areas of family law and company law.

Insufficient Questioning Regarding the Setting up of a Legal Entity

The notary plays a prominent role, especially during the phase that precedes the establishment of a legal entity. He provides the client with advice and fiscal information regarding the various types of companies that may be established. As he is trained in company law, he has the judicial knowledge necessary to deal with all aspects of the subject. The notary is able to choose the most suitable form of company for an entrepreneur's business plan. As mentioned above, the notary is impartial and has a legal obligation to provide the requested services. He is not entitled to turn down the request of a client without good reason. One manner of discovering the client's intentions is to keep on asking questions on the goal of the requested services. The cases and interviews indicate that notaries sometimes are too trusting with regard to the (twisted) reality presented to them by one of the parties involved. They are not sufficiently critical, and continue to render services despite their "gut feeling" that something is wrong.

> A striking example is the Dutch case of a notary who regularly established legal entities at the request of a major fraudster and also conducted share transactions on his behalf. Despite the fact that, given the dubious circumstances surrounding the establishments, the notary became suspicious, he allowed himself to be bamboozled on the basis of false information and continued to render services without further question. At one point, given the dubious circumstances surrounding the transactions, the notary consulted with a colleague who had rendered services to the client previously. Although they were not able to discover anything suspicious, the notary was left with a "gut-feeling" that his services were being abused. The notary did not conduct any deeper investigation into the background of his client and allowed himself to be misled on the basis of the documents. During the police interrogation, the prime suspect stated that he used he services of the notary because "he worked fast and did not ask tricky questions."

An Italian case that may help to illustrate the potential vulnerability of the notarial profession is a case concerning a "Ndrangheta" family in Lombardy which operated on the narcotics market.

> A notary, chosen by the organized criminal group, was given the task of managing the family's patrimony and established a financial holding in a foreign country. He then transferred to this holding all the shares of Italian companies, directly controlled by the criminal group and registered the shares in the names of relatives, friends and affiliates. From the financial holding he passed the shares to an ad hoc foreign company, and after some time the notary transferred all the shares to a series of straw men. These were subject to the control of the notary (through pledges and irrevocable mandatory powers), who was the sole administrator of the company. In this way, the straw men appeared to be the owners of the shares, but they could only use them to carry out operations wanted by the criminal group or the notary.

Lack of Due Care in Executing Deeds with Regard to Transactions on the Property Market

The purchase and sale of real estate property are activities during which notaries could be faced with money laundering operations. The property market is a well-known favourable place for criminals to invest their illegally obtained income. As the interference of notaries in this kind of activity is obligatory, inevitably he will also be confronted with dubious transactions.

Like the first manifestation of culpable involvement, the facilities of notaries who are not sufficiently critical and who do not ask any questions in odd circumstances, are more likely to be abused than the facilities of colleagues who do not take everything for granted. In the Dutch report references are made in this context to so-called *A-B-C-constructions*. This is a construction whereby a deed of conveyance for real estate is executed a number of times by one or more notaries within a short period of time (sometimes a single day). What is striking here is the fact that the value of the property rises steeply on each occasion. *A-B-C-constructions* are regular features in the property market. Such a construction in itself by no means always implies money laundering. However, in the event of several contracts of purchase and sale on the same day, the notary should certainly be alert and ask extra questions.

Sometimes, the activities of the notary can be regarded as culpable involvement in the strict sense.

> In the French report, a case is described of a notary who was visited by a well-known drug dealer. The latter wanted to buy a luxury apartment. Although the notary was well aware of the identity of the drug dealer, he did not object when the drug dealer proposed to use a false name on the deeds. After the notary was informed of the arrest of the drug dealer, he contacted his concubine and arranged that officially she was the one who had purchased the apartment. He also advised the concubine how to make the financial arrangements concerning the transaction as little transparent as possible.

Misrepresentation by Maintaining Deceptive Appearances

Like lawyers, notaries can also abuse their status, for example by helping to shield criminal activities or by shielding the proceeds thereof. For many people, the simple fact that a notary is present at a project meeting is interpreted as a sign that the project is reliable.

> A Dutch case refers to a notary who linked his name and account to an advanced fee fraud. As a result, a swindler was able to persuade investors to transfer huge sums of money to the account.
>
> In another Dutch case property was fictitiously divided up by a notary's deed into several apartments. A real estate agent sold all apartments to a straw man. Subsequently, each apartment was mortgaged for a sum far in excess of its value. False sales agreements, employer's declarations, salary slips and valuation reports were created for the mortgage application. The false documents and the valuation reports were drafted by the real estate agent. The same notary executed all deeds of transfer of title and mortgage applications. In several deeds of transfer, the notary amended the purchase price. On the basis of these fraudulent contracts, the financial institutions granted mortgages. The police file indicates that the financial institutions viewed the notary as a "guard" for their interests. They were confident that the documents received from the notary met the required standards and were drafted truthfully; hence they saw no need to carry out a final check.
>
> In Italy, a case was revealed in which a notary had falsely certified that the signatures of the vendors of real estate had been made in his presence.
>
> Another Italian case refers to a situation where a notary authenticated as true and having been affixed in his presence, the signatures of foreign citizens on labour contracts, which were then utilised to request the issue of residence permits.

Third Party Account as Safe-Haven

Similar to the system in place for lawyers, notaries in Italy, France and the Netherlands are obliged to use third party accounts. Notaries' third party accounts are equally susceptible to improper use, as shown by the following Dutch case.

In the Netherlands, a notary executed mortgage acts while he was aware, or should have been aware, that the institutions granting the mortgages were being misled. Furthermore, he amended concept deeds in such a way, that the amounts agreed were consistent with fraudulent contracts of sale. The profit generated by these fraudulent acts was booked to a co-suspect via the notary's third-party account.

Supervision of Lawyers and Notaries

The fact that lawyers and notaries occasionally exceed the limit in providing services to clients and become culpably involved in (organized) crime has prompted both governments and the professional groups to focus on counter-strategies. For a long time, the professional groups trivialised the problem by pointing at the low number of professionals convicted for culpable involvement in criminal activities. During the last decade however, the representative bodies of the lawyers and the notaries have definitely become more aware of the necessity to pay more attention to the issue of professional integrity.

In the context of supervision and control of lawyers and notaries, three levels can be distinguished. Firstly, there is the micro level of the firm employing the service-provider in question. There may be a lack of social control among lawyers and notaries in one-man firms, which, in other firms could hinder the development of undesirable activities. There is no accountability to colleagues or partners. Even so, it is still the case that large firms with strongly specialised and hence independent sections may also lack social control. Moreover, social control does not always mean that this control stands in the way of types of services being provided to organized crime in a manner that outsiders would regard as culpable or unlawful.

The second level is the meso level of the professional group. There are various forms of internal supervision within the professional groups: social control by fellow members of the profession, codes of conduct, statutes and directives, and, self evidently, disciplinary measures. In several countries, the national associations of lawyers and notaries introduced new guidelines, according to which their members have an obligation to examine the purpose of the request of a client, in case they suspect that their services might be abused. However, the extent of this obligation to examine remains unclear. Levels of *visible* enforcement of these provisions have been extremely modest in most countries.

Lastly, at macro level, the activities of lawyers and notaries are supervised from outside the professions. Where rendering services to organized crime are concerned, alongside informal supervision by the media and public opinion, it is above all a question of enforcement of the criminal law by the police and justice authorities, plus, since 1 June 2003, mandatory reporting on the basis of the Money Laundering Directive 2001.

As mentioned before, due to various reasons, the police and justice authorities do not devote much effort to investigate the culpable involvement of lawyers and notaries in organized crime. The question is what the effects are of the new obligation for lawyers and notaries to report unusual or suspect transactions. At this stage, the answer to this question is merely speculative, as in many countries the new Directive of the European Commission was admitted in national legislation quite recently. What we do know, however, is that so far the way law enforcement agencies have used information on suspicious financial transactions is not very promising.[10] Only a very small percentage of the suspicious reports is actually being used. There is no evidence whatsoever that the extension of the obligation to report certain financial transactions to other professional groups will lead to more – and more successful – law enforcement activities.

Concluding Remarks

Research shows that the evidence of culpable involvement of lawyers and notaries in organized crime cases comes from occasional cases, rather than any clear empirical trend or statistical basis. Given the nature of the problem, this is not surprising. After all, social scientists reproduce the blind spots of law enforcement officials, as they face the same problem: the privileges of confidentiality and non-disclosure itself operate as a very effective barrier to initial enquiry, which can only be overcome by strong evidence from other sources.

With regard to lawyers, the most striking evidence in the United Kingdom comes from systematic mortgage frauds and the huge sums of money transferred through solicitors firms in the course of high-yield investment frauds. In the other three countries, the case studies concerning lawyers predominantly refer to the assistance in various forms of money laundering schemes, and – to a lesser extent – to passing through confidential information to unauthorised persons, and the intimidation of witnesses.

Cases of culpable involvement of notaries mainly involve transactions on the property market, the establishment of legal entities to shield criminal activities, and drafting fraudulent deeds to enable money laundering. In general, rather than being considered as perpetrators or accomplices, the notaries involved in this sort of activity take an insufficiently critical stance with regard to the facts as presented to them.

It is probably not possible to prevent the abuse of lawyers and notaries by criminals entirely because the latter will, if they are prepared to take the time and expend the cost, often create complex transactions which appear legitimate, possibly even to the lawyer and notary. Thus, there is a limit to the extent to which professionals can effectively inquire into the propriety of their client's affairs.

---

[10] See Levi (2002) for the situation in the Anglo-Saxon countries. Research in the Netherlands by Terlouw and Aron (1996) showed that during the first years, only 3% of the reported *unusual* transactions were used for law enforcement purposes. The number of reports has accelerated since, but the capacity of the authorities to process them has not kept pace (Faber & Nunen, 2004).

Taking the contemporary developments within and outside the professional groups into account that may influence professional standards and may increase the risk for lawyers or notaries to become culpably involved in (serious) forms of crime, professional integrity increasingly will be discussed, both within and outside the legal profession. The growing awareness of the representative bodies and the introduction of several preventive measures may not only be helpful to individual lawyers and notaries who are facing a complex dilemma, but may stimulate the debate on professional ethics within the professional organisation in general as well. After all, the best way to keep the integrity of professional service providers at a high level is that lawyers and notaries themselves discuss openly the contemporary dilemmas they are facing.

References

Bunt, H. G. van de. (1996). Inzake Opsporing Bijlage X. Deel III onderzoeksgroep Fijnaut. *Beroepsgroepen en fraude*. 's-Gravenhage: Sdu uitgevers.

Chevrier, E. (2004). The French government's will to fight organized crime and clean up the legal professions; the awkward compromise between professional secrecy and mandatory reporting. *Crime, Law and Social Change, 42*(2–3), 189–200.

Faber, W., & van Nunen, A. A. A. (2004). *Uit onverdachte bron. Evaluatie van de keten van ongebruikelijke transacties*. Onderzoek en beleid 218. Den Haag: Ministerie van Justitie, WODC.

Fijnaut, C., Bovenkerk, F., Bruinsma, G., & Bunt, H. G. van de. (1998). *Organized crime in the Netherlands*. The Hague: Kluwer Law International.

Green, G. S. (1990). *Occupational crime*. Chicago: Nelson-Hall.

Kleemans, E. R., Berg, E. A. I. M. van den, & Bunt, H. G. van de. (1998). *Georganiseerde criminaliteit in Nederland, Rapportage op basis van de WODC-monitor*. Onderzoek en beleid 198. Den Haag: Ministerie van Justitie, WODC.

Kleemans, E. R., Brienen, M. E. I., & Bunt, H. G. van de. (2002). *Georganiseerde criminaliteit in Nederland, Tweede rapportage op basis van de WODC-monitor*. Onderzoek en beleid 198. Den Haag: Ministerie van Justitie, WODC.

Lankhorst, F., & Nelen, H. (2004). *Professionele dienstverlening en georganiseerde criminaliteit; Hedendaagse integriteitsdilemma's van advocaten en notarissen*. Zeist: Uitgeverij Kerkebosch b.v.

Levi, M. (2002). Money laundering and its regulations. *The Annals of the American Academy, AAPSS, 2002, 582*, 181–194.

Middleton, J., & Levi, M. (2004). The role of solicitors in facilitating 'Organized Crime': Situational crime opportunities and their regulation. *Crime, Law and Social Change, 42*(2–3), 123–161.

Nicola, A. di, & Zoffi, P. (2004). Italian lawyers and criminal clients: Risks and countermeasures. *Crime, Law and Social Change, 42*(2–3), 201–225.

Terlouw, G. J., & Aron, U. (1996). *Twee jaar MOT. Een evaluatie van de uitvoering van de wet melding ongebruikelijke transacties*. Onderzoek en beleid 158. Arnhem: Gouda Quint.

Law Enforcement

Tony Soprano: *"I'm not getting any satisfaction from my work, either"*
Dr. Jennifer Melfi: *"Why?"*
Tony: *"All because of Rico"*
Dr. Melfi: *"Is he your brother?"*
Tony: *"No, the RICO statutes."*
          (Sopranos)

Containment and Prevention

Chapter 11
Criminal Conflicts and Collective Violence: Biker-Related Account Settlements in Quebec, 1994–2001

Carlo Morselli, Dave Tanguay, and Anne-Marie Labalette

In the absence of channels for legitimate contract resolution, violence emerges as an alternative regulatory mechanism between conflicting parties (Black, 1998, 1983). In criminal settings, forms of regulatory violence have been referred to in a number of ways (e.g., contract killings, criminal liquidations, or gangland killings). We use the term "account settlement" to represent lethal or violent conflict resolution methods in criminal settings.[1] In this study, we examine fluctuation in an account settlement trend. Our empirical focus is exclusive to the Quebec context, in which account settlements represented approximately 17% of all homicides between 1985 and 1989 (Cusson, Beaulieu, & Cusson, 2003: 300).

In Quebec, increases in account settlements traditionally coincided with increases in overall homicide trends. However, since 1991, a unique pattern emerged. While a striking drop took place in the overall homicide rate, falling under two homicides per 100,000 for the majority of the years between 1991 and 2003, account settlements increased within this overall decline. A similar trend was also documented by Decker and Curry (2002) in their observations for St. Louis and other Midwest American cities. Although account settlements do increase from time to time in cyclical patterns that are often sustained for a number of consecutive months and, at times, years, the main reason for the increase in the Quebec context during this period was a prolonged biker conflict that took place between 1994 and 2001. This conflict is at the centre of our present analysis. By converging on the Quebec biker conflict, we find ourselves in the midst of a prolonged peak in an account settlement trend. Our principal analytical paths emerge from research that emphasizes shifts from individual to collective violence in high conflict settings.

Violence as Collective Behaviour

Assessments of violence in criminal enterprise and gang settings have generally followed explanations centering on subcultural characteristics (Klein & Maxson, 1989; Sullivan, 1989), violence-prone leaders (Yablonsky, 1959), and monopolistic

---

[1] The notion is also consistent with Tremblay and Paré's (2003) reference to co-offender homicides.

D. Siegel and H. Nelen (eds.), *Organized Crime: Culture, Markets and Policies.*   145
© Springer 2008

pursuits of market control or territorial disputation (Cressey, 1969; Schelling, 1984). A basic limit with such approaches is that these principal features must vary in accordance with account settlement patterns in order to establish causal inference. Hence, if more intensive cycles in account settlements emerge, we must be able to demonstrate that: violence is increasingly integrated within the subculture; a gang leader is increasingly aggressive toward other gangs; one group's strife for dominance is more evident than ever; or turf-warfare is more rampant. Each element must emerge more prominently than it would during a normal, stable period in order to capture cyclical patterns in a trend. Although there is strong reason to believe that each feature is likely present at any given moment in criminal enterprise settings, variations in their intensity have yet to be accounted for to any serious extent.

Probably the most common explanation that emerges in times of increasing account settlements is that of the aggressive and dominating gang leader. This is precisely what was expressed in Quebec. Police, journalists, and court officials were all in agreement that the biker conflict was caused by the reckless ambitions of one of the top Hells Angels in the province. Such reasoning is difficult to contest, but there is at least one study that demonstrates why a "gang war" is not likely to be ordered from the higher ranks of a criminal hierarchy. Levitt and Venkatesh (2000) studied the financial stucture of a crack-dealing gang and found violence within and between gangs to be a very costly experience. Only low-level foot soldiers benefitted from a context of lethal violence: in times of high risk, the salaries of lower level gang members increased, while the earnings of higher ranking members and overall gang revenues decreased substantially. The authors raised the agency problem in explaining this concentration of financial benefits. They suggested that lower level members are more likely to be responsible for the rise of violence in a given setting: "Much of the violence is not sanctioned by the gang, but rather arises because a particular foot soldier interested in moving up the hierarchy may have a strong incentive to build a reputation for toughness and thus may engage in violence even if such actions run counter to the best interests of the gang. Once such violence occurs, it is difficult for the opposing gang not to retaliate" (Levitt & Venkatesh, 2000: 781).

Decker (1996), who studied intergang violence within the more general collective behaviour framework suggested by Thrasher (1927), described what happens when retaliation is triggered. He maintained that violence is an integral part of the gang environment, but he also provided an explanation for an increase in violence during certain periods: "rapid escalation of intergang hostilities" is a function of retaliation or, in Loftin's (1984) terms, contagion. "Such actions reflect the collective behaviour processes at work, in which acts of violence against the gang serve as the catalyst that brings together subgroups within the gang and unites them against a common enemy. ... The precipitation of such activities pulls fringe members into the gang and increases cohesion" (Decker, 1996: 256).

Decker's focus offers another frame of reference from within the gang, where solidarity is intensified in times of strife. However, we know little about how the level of group solidarity intertwines with the wider criminal milieu that serves as a backdrop for interactions between groups or individual participants. An increase in solidarity within the gang is a reaction to changes in the larger milieu. This is supported by research on conflict beyond gang and organized crime settings.

Conflicting Parties, Group Action, and Social Polarization

Although not directly concerned with account settlements, Gould's (2003) research on the interplay between various forms of conflict and violence provides a conceptual scheme to begin building our own analytical framework. Gould studied kinship-based conflicts and vendettas in ninteenth century Corsica and exposed the internal mechanisms that either deter or escalate collective violence between conflicting parties. Collective violence has been a key concept in research on riots, protests, and crowd violence. Gould's application also comprised the ensemble of more prolonged and scattered incidents of violence. In his (and our) framework, collective violence represents the onset and rise of significant group mobilization and solidarity around a violent conflict (Gould, 2003: 112–115).

Gould's general statement maintained that although violence emerges as a suitable conflict-management alternative in resolving a dispute between individuals, repeated retaliation and escalation in violence occur primarily because at least one party in the conflict mobilized (or, took the form of) a group in settling the dispute violently. Two important nuances are made in elaborating this argument.

First, Gould stressed the rarity of collective violence even in settings known to be prone to violence. As with the Corsican context, criminal milieus are often assumed to be marked by high levels of violence and subsequent retaliations. But considering the unlawful circumstances within which criminal entrepreneurs generally find themselves, it would be more reasonable to ask why violent resolution methods are not more frequent.

Second, Gould (2003: 115) illustrated that manifestations of group solidarity underly both deterrence and escalation of group conflict: "Expressions of group solidarity are therefore double-edged: they may succeed in forestalling escalation, but if they fail to do so they intensify violence that occurs." Although individuals are likely to associate with others in partial or established associations, group affiliation serves to deter potential clashes, but also intensifies retaliation once it is triggered. In the Corsican setting, retaliations generally took place when victims were attacked by groups. Conflicts between lone individuals (although group members) were rarely met with retaliation. Vengeance is therefore presented as a sanction against group solidarity or collective aggression within the conflict and the advent of collective violence is the progressive outcome of the onset and escalation of group-based retaliation.

Gould's argument helps us interpret cyclical increases and fluctuations in account settlements. During periods of normal or stable account settlement trends, conflicts are more likely taking place between individuals, regardless of their group membership. In times of abnormally high account settlement trends, such conflicts are more likely taking place between groups who increasingly emerge as active units around their individual members involved in conflictual situations. This is the first sign of a shift from individual to collective violence.

Hence, there is a basic difference between *group interests* and *group actions*. Normal account settlement levels indicate that group-based retaliations are generally not taking place and that violence is occurring primarily between individuals. Although any individual involved in a conflict may be part of a group in his business operations (the presence of group interests), managing a conflict does not incorporate the active presence of that group (the absence of group action). Abnormal increases in account settlements indicate an initial group presence within any given conflict. The mobilization of a group to support one party within a conflict will be met with a mobilization effort to support the other party (the presence of group action). The enhancement of group identification (or solidarity) brings collective liability to the forefront, as the group is held accountable for the actions of its individual members. Thus, in shifts from individual to collective violence, violence is not only performed by and for the group, it is also perceived by others as a group effort. Within this transition towards group action and collective violence, the criminal milieu becomes increasingly polarized. In addition to intensified group identification, other participants who would normally remain independent from either conflicting group often choose, or are required to choose an allegiance.

Thus, the application of a collective violence framework to the criminal milieu suggests that what is often referred to as an underworld or criminal gang "war" is more likely a chain-like series of vengeful acts (Cusson, 1989: 104). Without any form of intervention either from within or beyond the criminal setting, a tit-for-tat retaliatory process between conflicting parties inevitably emerges. We now turn to the Quebec biker conflict to illustrate such a process.

The Biker Context in Quebec

A distinction is made between non-criminalized and criminalized (or outlaw) bikers (Alain, 2003; Tremblay et al., 1989; Wolf, 1991). The latter are generally qualified by motocycle club members as "one-percenters" to refer to those clubs that are not registered with the American Motorcycle Association or the Canadian Motorcycle Association (see Barger, 2000; Wolf, 1991). Others have reduced the one-percenter classification in exclusive reference to the "Big Four" clubs (e.g., the Hells Angels, the Outlaws, the Bandidos, and the Pagans) (see Quinn & Koch, 2003). In Quebec, as elsewhere, many motorcycle aficionados are active members of several clubs, but only a negligible proportion of this population has been the central concern of law-enforcement controls throughout the past three decades. Groups, such as the Hells Angels, that make up this minority have traditionally been associated with a simultaneous existence as a criminal enterprise. Our focus is on participants either within or associated to such groups.

Alain (2003) and Tremblay et al. (1989) scanned the recent history of one-percenter biker groups in Quebec and identified three phases. In the first phase, from 1971 to 1978, most of this population was located in rural Quebec and maintained a population of roughly 600 individuals in its early days and climbing to 900 members towards the end. During this period, many alliances were formed between small groups. This period was marked by the arrival, in 1977, of the first Hells Angels chapter (in Sorel, Quebec) who established their presence as an over-riding organization. During the second phase, from 1978 to 1983, roughly 800 individuals were identified within the one-percenter classification. This period was marked by increased violence between clubs and by the positioning of several groups within various illicit drug markets (particularly for synthetic drugs). According to Alain (2003), the number of one-percenter bikers dropped to 300 during the final phase from 1984 to 2001. This period also highlights the presence of the Hells Angels as a dominant group.

Although we agree with this assessment of the one-percenter biker population in Quebec, we chose to divide this last period in order to adequately expose the biker conflict at the centre of our analysis. We establish a clear change since 1994. The predominant clubs were the Hells Angels and a Montreal-based group, the Rock Machine.[2] The Rock Machine was at the core of an amalgam of small biker groups and independent drug merchants who joined forces to form the Alliance during the latter half of the 1990s in Quebec. The biker conflict that significantly contributed to account settlements was played out between the members and affiliates of the Hells Angels and the Alliance.

Between 1994 and 2001, a substantial number of killings took place within this ongoing conflict in Quebec. Previous periods (most notably the late 1970s and mid-1980s) had been punctuated by important events and atypical levels in biker-related homicides, but no other period in Quebec or Canadian history has been marked by such consistent and clustered account settlements over such an extended time. This conflict also had the disctinctive feature of reaching beyond the biker or criminal milieu. Several incidents were critical for understanding how the conflict evolved.

Source for Account Settlements

Our case study is based on a data set of the biker conflict under analysis. The data were gathered primarily in criminal intelligence files and include all account settlements

---

[2] Initially, the Rock Machine was not a biker club. In fact, the evolution of the biker club/criminal enterprise hybrid took place inversely to the experiences of most other one-percenter groups. While most groups shift from biker club to criminal enterprise (at least for some of its members), the Rock Machine began as a criminal enterprise and shifted gradually into a biker club. As the Quebec-based conflict with the Hells Angels grew, the identity of the Rock Machine became more biker-oriented. Becoming a Bandidos outpost and official Bandidos chapter marks the final phase in this unique transition.

suspected of occuring during the escalating conflict between members and affiliates of the Hells Angels and the Alliance in Quebec. Until now, the term account settlements has been used to refer to crime-related conflicts that were regulated by murder. For operational purposes, we widen the scope of this reference to include attempted murder as well.

Two people, both working for the Sûreté du Québec (Quebec's Provincial Police), were responsible for keeping track of the increasing number of account settlements between biker groups. One was a senior investigator reputed as the foremost law-enforcement expert on outlaw biker matters – he was a central figure in gathering information on biker club members and their network participants. The other was a strategic analyst – she was responsible for validating the investigator's information and for continually updating the present data set. Construction of the data set ended when an immense crackdown took place on March 28, 2001 (known as Springtime Operation 2001), in which 125 suspected members and affiliates of the Hells Angels were arrested. Soon after this major event, they provided us with a copy of this final data set.

The inclusion of each account settlement case is questionable because, prior to March 2001, not a single person had been officially accused of any of these acts of lethal violence. At the start, the data set included information pertaining to the date, setting, and method used, as well as the victim's name, age, group affiliation, and club status. For our own purposes, an account settlement corresponds to the targeting of an individual – our cases are victim-based, rather than event-based. We were concerned with the validity of three types of information surrounding such acts: (1) that the event situating the account settlements did take place (in the case of attempted homicides); (2) that it was indeed biker related; (3) that the gang affiliation of each victim was accurate.

Overall, 361 victims (174 murders, 174 attempted murders, and 13 disappearances) were initially recorded in this data set. Our verification and focus of analysis led to the exclusion of several cases. First, we were interested in analyzing the conflict between participants in the biker milieu. Hence, 20 individuals who were either innocent bystanders in a crossfire situation or who were targeted by either group for reasons indirectly related to the conflict (e.g., two prison guards, a journalist) were excluded. Second, for the purpose of our analysis, 13 cases classified as disappearances were also removed from the initial data set due to their ambiguous status within law-enforcement circles. Third, 67 victims were removed because their affiliation to either side of the conflicting groups could not be established and confirmed with newspaper sources.

Our verification reduced the data set to 261 victims – specifically, 126 murders and 135 attempted murders that were classified and confirmed as having taken place within the Hells Angels/Alliance biker conflict between 1994 and 2001 in Quebec. Of the 261 account settlements, 55.2% were aimed at Hells Angels members or affiliates, and 44.8% were aimed at participants in the Alliance. We aggregated this data on a monthly basis ($n = 96$ months).

Sources for Other Components in the Biker Conflict

Aside from having an impact on the province's criminal milieu, the length and intensity of the the biker conflict created overriding concerns among the general public and within law-enforcement settings. Three elements of these widening implications are carried over into our analyses: (1) public sensitizing events; (2) implementation of specialized investigative squads designed to target the biker milieu; (3) arrests of biker members and associates.

Public Sensitizing Events

In the past, rampant criminal conflicts have often led to collateral damage in the general public. For example, the zeal that accompanied the advent of state control of suspected mafia operations in Sicily and other regions throughout Italy during the 1980s was preceded by a series of killings that involved innocent or non-criminal victims (e.g., innocent onlookers, journalists, judges, police officials, local politicians). Such violence was key to sensitizing a large portion of the public that had previously been silent toward the growing problem (see Jamieson, 2000). Subsequent public reaction oriented the scope of state intervention against the implicated criminal group(s).

Although a substantial number of such incidents also occurred in relation to the Quebec biker conflict, we restricted our focus to four events considered as the most influential based on the media attention they received. The first, and likely the most important event, took place in August 1995 when a young boy was accidentally killed by debris from the exploding vehicle of a Hells Angels affiliate (who was also killed). The second and third events were the murders of two prison guards in June and September 1997. The fourth event happened late in the conflict in September 2000. A long-time Montreal crime journalist was shot the day after publishing a controversial article speculating on various suspects in the deaths of a number of victims in the conflict. These events are expected to have a direct impact on the evolution of account settlements, as well as an indirect influence on the law-enforcement response to the conflict.

Law-Enforcement Phases

Two months after the boy's accidental death, a specialized investigative squad was assembled to target participants in the biker milieu and in the emerging conflict. This particular law-enforcement attention took place along three phases. In October 1995, the Montreal Wolverine squad was put into operation. This squad amalgamated the efforts of investigators from the Sûreté du Québec, the Montreal Police, and the Royal Canadian Mounted Police. Their specific focus on the Montreal region ended during the summer of 1996, when the Wolverine squad expanded to include the region of Quebec City. The Montreal/Quebec Wolverine squad was active for another two years until May 1998, when it was decentralized and replaced with a province-wide system of six investigative squads. This last group of Mixed Regional Teams marked the final law-enforcement phase concerned with the conflict.

Arrests

The number of arrests of each group's members and associates is one straightforward indicator for measuring the level of law-enforcement intervention throughout the conflict. Although various police agencies denied us access to information pertaining to the specific numbers of arrests, we gathered such information from six newspaper sources across the province during the conflict period. Overall, 1089 arrests were recorded. The majority involved suspected Hells Angels affiliates (86.4%). Notwithstanding the fact that various police agencies across the province were likely to be more focused on targeting this established and more notorious group, two other reasons account for this disproportionate attention paid to Hells Angels affiliates. First, the Hells Angels was the larger group and its affiliations across the province were broader than those of the growing Alliance consortium. Second, our estimate includes the 125 arrests of Hells Angels affiliates made at the end of March 2001.

As with the account settlements, some of these arrests involved the same individuals. Again, this was compatible with the focus of our study because we were primarily interested in accounting for the level of pressure exerted on the biker milieu at any given time. All arrests were categorized on a monthly basis.

Assessing the Biker Conflict Within the Account Settlement Trend

The first objective of our analysis is to establish increases in general account settlements. Figure 11.1 illustrates the fluctuating proportion of account settlements between 1988 and 2002. Although signs of increased killings are revealed as early as 1990 and 1991, the smoothed curve highlights the rise during our case study's time frame between 1994 and 2001.

The increase becomes more prominent if we consider the presence of gang-related incidents from 1994 to 2001. Using data from Statistics Canada's Homicide Survey, a rapid increase is observed as of 1994 and continues until the end of the conflict period in 2000. Account settlements begin to drop as of 2001 (see Fig. 11.2).

The representation of gang-related account settlements is consistent with Cordeau's (1990) analysis of account settlements in Quebec between 1970 and 1986. Peak years for overall account settlements (1973–1976 and 1983–1985) also had a higher proportion of group-related conflicts that persisted over relatively longer periods. They were also marked by a series of retaliatory killings between conflicting groups, particularly within the Montreal region (see Cordeau, 1990: 51–64). Closer analyses of both peak periods revealed that many of these

(Quebec, 1988-2002)

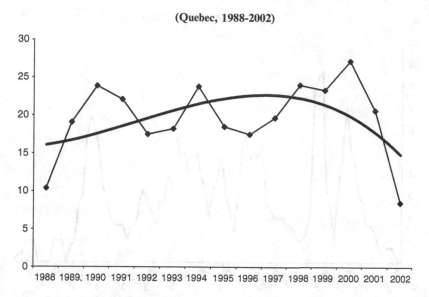

Fig. 11.1 Percentage of account settlements per all homicides. *Sources*: (1) Account settlements (Allo Police); (2) All homicides (Homicide Survey, Statistics Canada)

(Quebec, 1988-2002)

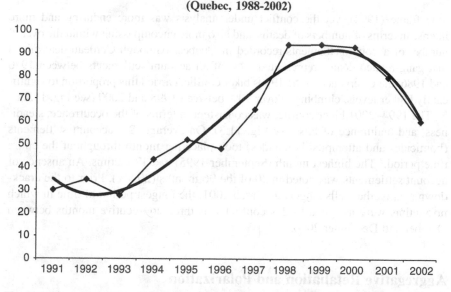

Fig. 11.2 Percentage of gang-related account settlements per all Account Settlements. *Sources*: (1) All account settlements (Allo Police); (2) Gang-related account settlements (Homicide Survey, Statistics Canada)

gang-related account settlements leading to such increases were due to biker-related conflicts (see Cordeau, 1990: 64–67).

The approximate 10-year span separating these previous peak periods (1973–1976 and 1983–1985) is repeated between the last peak and the starting point of our study's

(3-month moving average)

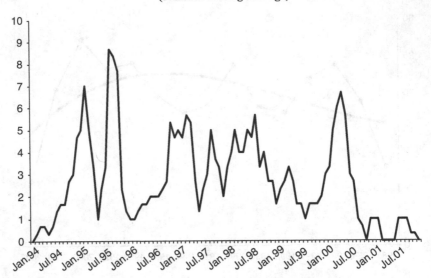

Fig. 11.3 Account settlements attributed to biker conflict, 1994–2001

time frame (1994), yet the conflict under analysis was more enduring and more intense in terms of numbers of deaths, and also more encompassing within the overall number of account settlements recorded in Quebec. Although Cordeau determined that gang-related deaths represented 19% of all account settlements between 1970 and 1986, the emergence of the 1990s biker conflict carried this proportion to significantly higher levels, climbing to over 90% between 1998 and 2000 (see Fig. 11.2).

The 1994–2001 biker conflict was consistent in terms of the occurrence, steadiness, and endurance of hits (see Fig. 11.3). On average, 2.7 account settlements (homicides and attempted homicides) took place per month throughout the entire time period.[3] The highest month (September 1995) had 17 victims. An absence of account settlements was noted in 20 of the 96 months (20.8%). Prior to the crackdown against the Hells Angels in March 2001, the longest period of time in which no victims were assigned to this conflict was three consecutive months between October and December 2000.

Aggregative Retaliation and Polarization

Using Gould's (2003) terminology, our biker conflict may be indicative of mere collaborative violence, marked by several short-term conflicts between members of groups, yet never reaching the level of generalized violence that grasps an increasing

---

[3] Although Fig. 11.3 is presented with a 3-month moving average, the facts surrounding the account settlement trends in this analysis are based on monthly trends.

number of independent participants within the escalating conflict. For example, in Quebec during earlier periods, Cordeau (1990) described a chaotic pattern in gang-related conflicts that appeared to be both short-term and with ephemeral alliances (119: 63). Although observing account settlement peaks that surpassed 2-year periods, Cordeau's interpretation would not favour a generalized violence scenario spanning a lengthy period. Rather, what appears to be long-term and cohesive at the aggregate level is in fact an amalgamation of several independent scenes of collaborative violence taking place within the same time frame. Cordeau acknowledged that the criminal milieu was experiencing some form of transition, but, empirically, links between the ongoing conflicts were not attested.

Quite differently, the 1994–2001 conflict did grow into a context of generalized violence. Our analysis provides an illustration of the structure and concentration of the conflict. The conflict was not a series of disassociated smaller clashes. It was an ordered process in which allied and independent participants were increasingly forced to either realign or align themselves to one of the implicated biker gangs. It not only intensified throughout the period, but did so within a retaliatory and increasingly polarized pattern.

The creation of the Alliance was, in itself, an indication that independent and smaller groups of drug merchants were consolidating affiliations with the group that was growing around the Rock Machine. What had once been a drug distribution setting that was characterized by scattered groups and interacting independent participants, radically changed into a cohesive setting built primarily around two conflicting parties.

Our data set illustrates the collective violence that was in place during this time period. By distinguishing the affiliations of victims, we are able to see how account settlements against one group match those against the other group (see Fig. 11.3). We did this by using a cross-correlation technique. Cross-correlations are similar to Pearson correlations but with the time factor (e.g., a month) taken into consideration. This technique allows us to assess if and how two trends coincide at different intersections in time. By correlating the ensemble of hits against the Hells Angels with those against the Alliance, the hypothesis that a collective conflict was in place and that it was escalating in retaliatory fashion can be confirmed if the correlation appears at time 0 or, to account for the turning points of our selected time periods, within a 1-month time span. Cross-correlations at times −1, 0, or +1 indicate that both groups were experiencing common variations in account settlements within the same time period. This would substantiate that a single, generalized conflict was indeed in place. Failure to identify significant cross-correlations would leave us with an interpretation that is similar to Cordeau's, insofar as it would be difficult to defend both the partition of conflicting parties into two separate groups (hence, dispelling the likelihood of polarization within the criminal milieu) and the assessment that the conflict was indeed at a collective, generalized level.

Figure 11.4 indicates that the distribution of account settlements against the Hells Angels (top half) is strikingly symmetrical with the flow of account settlements against the Alliance (bottom half). This is true for both monthly and t3-month moving average trends. For monthly trends, cross-correlations emerge at time 0 ($r=0.24$; $\alpha<0.05$) and at time +1 ($r=0.26$; $\alpha<0.05$). For the 3-month moving average trends, cross-correlations are found at time −1 ($r=0.41$; $\alpha<0.001$), time 0 ($r=0.48$; $\alpha<0.001$),

(3-month moving average)

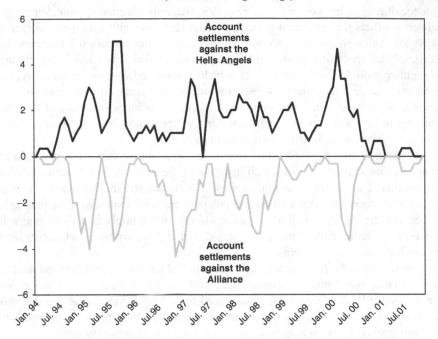

Fig. 11.4 Account settlements by biker gang

and time +1 ($r=0.43$; $\alpha<0.001$). The retaliatory pattern that emerges within such close time frames substantiates the biker conflict as a case of collective violence and also exposes the inherent polarization that takes place in such contexts.

Attrition, Public Implication, and the Law-Enforcement Response

The response of law-enforcement to the escalating conflict was marked by a series of phases that shaped the way participants on each side were targeted. Figure 11.5 superimposes the ensemble of law-enforcement phases (see top axis) and arrests across the 8-year conflict period.

Each of the public events aroused public awareness which subsequently affected the way that law-enforcement and other officials responded to the conflict. In terms of police response, little had taken place until the fall of 1995. It was only after the boy's death, in the summer of 1995, that the first Wolverine squad was convened and arrests increased. Although law-enforcement interventions continued intensively and consistently throughout the rest of the period, it took more than 3 years for this heightened control to have a notable impact on account settlements. However, we are not able to establish whether the decrease in account settlements and fluctuations throughout the entire conflict were due to these law-enforcement interventions.

(3-month moving averages)

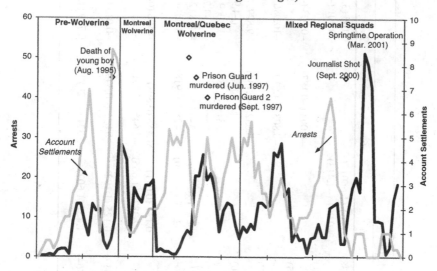

Fig. 11.5 Biker-related arrests, law-enforcement phases, and public events

It is possible that account settlements waned due to straightforward attrition amongst participants on both sides. After so much retaliation, the numerous deaths and prevailing violence resulted in fewer people left to target, particularly on the side of the smaller Alliance. This scenario can be developed by studying the impact of cumulative account settlements on monthly account settlement patterns. Such an analysis allows us to identify how monthly patterns are dependent on the number of account settlements that have already taken place. Whereas monthly incidents early on in the conflict should rise with a growth in the cumulative count (an escalation, rather than an attrition process), this relationship should inverse as the cumulative count rises and the conflict endures (the attrition process). Statistically, we capture this curvilinear pattern in two steps (see Model 1 in Table 11.1) First, we add a cumulative account settlement measure that is designed to identify the positive and linear relationship that marks the early months (a higher concentration of account settlements will amount to a monthly increase in forthcoming account settlements). Second, this same cumulative account settlement measure is squared in order to test whether later cumulative counts are inversely related with later monthly account settlements (once the cumulative count increases beyond a critical point, monthly account settlements drop). If an internal attrition process emerged within the conflict, the cumulative count impact should be positive (a gradual rise), and the impact of the squared cumulative count should be negative (an eventual drop). The combination of these two measures highlights the quadratic fit that best represents the impact of the growing conflict on account settlements at any given month, with the curve and drop being captured by the squaring of the initial variable. Table 11.1 indicates that this dual pattern is indeed present. After establishing this internal

Table 11.1 OLS regression of monthly account settlements (n = 96)

	Model						
	1	2	3	4	5	6	7
Cum. account Settlements	0.037 (2.517)*	—	—	0.143 (4.838)***	0.061 (1.660)	—	0.106 (3.185)**
Cum. account settlements^2	−0.001 (−2.832)**	—	—	−0.001 (−4.941)***	−0.001 (−1.413)	—	−0.001 (−2.266)*
Number of arrests	—	−0.009 (−0.602)	—	−0.010 (−0.695)	—	−0.003 (−0.187)	−0.003 (−0.244)
Montreal Wolverine	—	−1.917 (−1.344)	—	−7.391 (−4.330)***	—	−8.803 (−4.619)***	−10.086 (−5.405)***
Montreal/Quebec Wolverine	—	0.277 (0.258)	—	−8.221 (−3.870)***	—	−6.832 (−3.734)***	−9.829 (−4.998)***
Mixed regional squads	—	−1.028 (−1.079)	—	−10.086 (−3.517)***	—	−7.381 (−3.580)***	−12.069 (−4.766)***
Child's death	—	—	1.264 (1.362)	—	−1.871 (−.906)	8.234 (4.613)***	4.418 (2.152)*
First guard's death	—	—	−1.154 (−0.657)	—	−2.319 (−1.220)	−1.428 (−0.918)	−3.001 (−1.915)
Second guard's death	—	—	0.819 (0.477)	—	0.137 (0.071)	1.455 (0.898)	0.506 (0.319)
Shooting of journalist	—	—	−2.707 (−3.038)**	—	−2.382 (−1.665)	−2.645 (−3.526)***	−3.038 (−2.461)*
AR(1)	0.163 (1.589)	0.246 (2.282)*	0.115 (1.095)	0.156 (1.417)	0.125 (1.192)	−0.037 (−0.349)	−0.015 (−0.144)
Constant	1.592 (1.910)	3.367 (4.320)***	2.370 (3.474)***	0.640 (0.764)	1.427 (1.607)	2.371 (4.329)***	0.739 (1.010)

Slope (T-ratio)

*$p < 0.05$; **$p < 0.01$; ***$p < 0.001$

attrition effect, we continued to build our model with the addition of external components that influenced account settlement patterns during the conflict.[4]

By arresting and removing participants involved in the conflict, police likely accelerated this attrition process. The final crackdown (Springtime Operation) took place at the end of March 2001. Figure 11.5 shows that this event marked the end of the biker conflict. The law-enforcement response also had an impact beyond this ultimate and intensive control. Building on the initial internal attrition impact, law-enforcement controls and public sensitizing events are added in order to develop a fuller assessment of the factors that influence such conflicts (see Table 11.1). The number of arrests do not have an impact on account settlements. Whereas none of the specialized investigative squads emerged as significant components when studied independently of other factors (Model 2 in Table 11.1), their respective impact is substantially changed when we include either the internal attrition indicators (Model 4) or the public events that extended from the conflict (Model 6). In both models, each of the three phases of specialized investigative targeting significantly decreasd the monthly account settlements.

The significant impact of all phases of law-enforcement presence within the conflict is countered by the ineffectiveness of arrest actions that extended from these organizations. Why is police presence more salient than increasing arrests? First, arrests are likely contained within the scope of the phases' impacts. Second, arrests are indicative of specific actions against participants, but investigative phases are indicative of the type of overall resources allotted to law-enforcement during the conflict. In short, these phases offer an image of law-enforcement as an additional group in the ongoing conflict. Higher-order controls, therefore, emerge as better predictors than lower-level actions such as arrests.

Hence, heightened law-enforcement is a strong negative factor if we also take into account that internal checks were taking place within the biker milieu itself or that public awareness had been stimulated as a result of collateral damage (the boy's death and the shooting of the journalist). When considered independently from other factors, the attempted homicide of the journalist is the only public event that emerges as a significant (negative) predictor of account settlements (Model 3). This singular impact is lost when we consider the internal attrition effect (Model 5), but re-emerges when public events are combined with law-enforcement controls (Model 6). In Model 6, the boy's death also emerges as a significant factor. This positive effect could be due to the substantial number of account settlements that were recorded for the month following this event (recall that September 1995 had the highest number of incidents). It could also be due to the different impact that such collateral damage can have at early and later phases in the conflict. Model 6, although not controlling for internal attrition, is important because it is only after considering the boy's death and the shooting of the journalist that we note any significant negative impact on account settlements by various law-enforcement responses.

---

[4] The models in Table 11.1 all include an autoregression control (AR1). Aside from a slight impact in Model 2, all results are independent of time-series effects.

Model 7 is our complete model. The consistent impact of internal checks within the conflict itself is indicated by the positive effect of the cumulative count of account settlements and by the negative effect of the curvilinear extension. Specialized investigative phases continue to have important negative effects on account settlements. Finally, as in previous models, the same two public events continue to influence. Note that for all models except Model 2, the AR(1) parameter was not significant, suggesting that the temporal feature of the conflict was not, in itself, a factor. Instead, we find that each component was sensitive to the inclusion or removal of any other component. Internal attrition and law-enforcement containment are the two key elements for understanding how the biker conflict came to an end, yet both are sensitive to public events. Public events are key to understanding changes in internal attrition and law-enforcement controls, yet prove less telling when studied alone.

Conclusion and Extensions

Understanding interpersonal violence between participants in criminal settings requires an elaboration of the structural changes within which interactions take place. Although research on account settlements in criminal milieus has yet to pursue such a framework, studies of high conflict settings provide necessary insights on group action, collective liability, and social polarization. These concepts are central for understanding fluctuations in account settlements and general homicide trends.

Our findings indicate that the rise in account settlements in Quebec during the latter half of the 1990s was due largely to the increasing number of deaths resulting from a specific biker conflict. We also maintain that this conflict corresponds to a case of systematic and time-ordered retaliation between two clearly identifiable groups. The biker conflict was a case of collective violence over a prolonged period, and the criminal milieu in Quebec became polarized around these two groups. Sustained polarization and collective violence within this milieu had limits, but external intervention from specialized law-enforcement squads was required to bring the persisting conflict to an end.

Two important empirical limits persist in our exploration. First, we are not able to pursue how social polarization increases and sustains violence. This is a third party issue. Second, we are not able to assess the impact of external third party strategies on rising violence within such settings. These limits are elaborated in order to direct future research.

Polarization and Internal Third Parties

Social polarization within a high conflict settings reduces the number of available third parties who normally circumvent violence (Black, 1998, 1983; Cooney, 1998;

Cusson, 1999; Flap, 1988; Senechal de la Roche, 2001). Gould (2003) found a third-party effect in his own study but dismissed it as trivial. Understanding the prevalence and influence of third parties in criminal milieus is fundamental to grasping the polarization that emerges from peaks in account settlement trends. Criminal settings rely considerably on the interventions of informal third parties. Formal and external third parties, such as police or judicial agents, do not normally intervene in conflicts between criminal entrepreneurs. Informal arbitrators have been deemed intrinsic to various criminal settings – they occupy a role similar to their formal and external counterparts. This is corroborated by numerous studies of Mafia-like groups and wider patron-client systems (Albini, 1971; Anderson, 1979; Blok, 1974; Gambetta, 1993; Haller, 1991; Hess, 1998; Ianni, 1972; Reuter, 1983; Varese, 2001). Naylor (1997: 15), in studying the criminal marketplace, is the most succinct on this matter: "Contrary to the ideological fulmination of free-market economists, business needs government, and where government does not exist, it will be invented. That applies not just to legitimate society butn also to the under-world where parallel governments perform several functions."

Our suggestion to pursue third party intervention as a critical factor does not pertain to the influence that such actors have in resolving dyadic conflicts, but rather has to do with what happens to such actors when group action and collective violence is triggered. If the result is a reduction of internal third parties, a conflict will lack regulators and is therefore likely to escalate. Black's repressive peacemaker (the police) then becomes the lone intervening third party. Although law-enforcement intervention in normal periods of account settlements is minimal, its presence as an external third party becomes an overriding condition for ending or containing collective violence, hence, bringing the account settlement trend back to a normal level.

Such conflicts nonetheless contain their own checks (attrition) regardless of the lack of internal mediators. Yet, as in many extended periods of collective violence, such checks occur only because opposing parties are largely permitted to determine their own boundaries of violence. Our finding that law-enforcement controls (independent of internal attrition or public events) do not significantly decrease account settlements confirms that participants in such criminal settings are left to their own devices. This results in extreme criminal conflicts that expand and persevere over extended periods. Collateral damage should be expected under such circumstances.

External Third Party Strategies

How may collective violence in criminal conflicts be reasonably observed and predicted? In the Quebec context, newspaper coverage of the increasing number of incidents indicated that the conflicting groups were clearly identifiable as early as spring 1995. Hence, it was well known that a violent conflict was escalating between two groups, but few people were aware that such polarization around two groups is abnormal, even in criminal settings. Although concentrated, short-term peaks in account settlements emerge from time to time, the persistence of a growing number of incidents involving members and associates of the same suspected groups over a series of consecutive months suggests that a largely unregulated conflict is collectivizing. What appears, at first, to be a growing conflict between two groups is, after closer analysis, a more serious reshaping of the overall milieu around this conflict. By this point, external intervention is necessary, unless a strategy is in place to let internal attrition follow its own course,.

Law-enforcement interventions, moreover, must be designed either around non-partisan or partisan lines. The repressive peacemaker represents a non-partisan police reaction whose main aim is to arrest or constrain as many participants as possible regardless of whatever side the participant is a member. Though this approach appears obvious, it is not necessarily more effective than partisan police approaches. The decision involves choosing to regulate either a criminal market or a criminal conflict. One partisan police strategy would maintain more intensive targeting of the dominant group in order to reduce its presence within a given criminal market while that group is implicated in an intensifying conflict. Such a strategy prioritizes the control of the criminal market over the control of the criminal conflict insofar as it reduces the dominant group's force within the criminal market without necessarily controlling the increasingly violent conflict in the most timely manner. The second partisan option would call for a stronger focus on the contending group. Although this could help sustain and possibly even enhance the dominant group's presence in the criminal market, it would also concentrate police resources on the weaker side of the conflict. In this strategy, control of the conflict rather than control of the criminal market becomes the priority. Removing the smaller group might be the more suitable option in the face of intensifying violence.

Acknowledgments We thank Geneviève Gilbert, Pierre Avon, Frédéric Lemieux, Pierre Tremblay, Maurice Cusson, Marc Ouimet, and Martin Bouchard for their help and suggestions throughout various phases of this research.

References

Alain, M. (2003). Les bandes de motards au Québec: la distinction entre crime organisé et criminels organisés. In M. Leblanc, M. Ouimet, & D. Szabo (Eds.), *Traité de criminologie empirique* (pp. 135–160). Montréal: Presses de l'Université de Montréal.

Albini, J. L. (1971). *The american mafia: Genesis of a legend*. New York: Meredith.

Anderson, A. G. (1979). *The business of organized crime: A Cosa Nostra family*. Stanford: Hoover Institution Press.

Barger, R. (2000). *Hell's angel*. New York: Harper Collins.

Black, D. (1983). Crime as Social Control. *American Sociological Review, 48*, 34–45.

Black, D. (1998). *The social structure of right and wrong* (revised ed.) San Diego: Academic Press.

Blok, A. (1974). *The mafia of a Sicilian village, 1860–1960: A study of violent peasant entrepreneurs*. New York: Harper and Row.

Cooney, M. (1998). *Warriors and peacemakers: How third parties shape violence*. New York: New York University Press.

Cordeau, G. (1990). *Les règlements de comptes dans le milieu criminel québécois de 1970 à 1986*. Doctoral dissertation, School of Criminology, Université de Montréal.

Cressey, D. R. (1969). *Theft of the nation: The structure and operations of organized crime.* New York: Harper University Press.

Cusson, M. (1989). Les zones urbaines criminelles, *Criminologie, 22,* 95–105.

Cusson, M. (1999). *Le tiers exclu et la non-violence* (unpublished paper).

Cusson, M., Beaulieu, N., & Cusson, F. (2003). Les homicides. In M. Leblanc, M. Ouimet, & D. Szabo (Eds.), *Traité de criminologie empirique* (pp. 281–331). Montréal: Presses de l'Université de Montréal.

Decker, S. H. (1996). Collective and normative features of gang violence. *Justice Quarterly, 13,* 243–264.

Decker, S. H., & Curry, D. G. (2002). Gangs, gang homicides, and gang loyalty: Organized crimes or disorganized criminals. *Journal of Criminal Justice, 30,* 343–352.

Flap, H. D. (1988). *Conflict, loyalty, and violence: The effects of social networks on behaviour.* Frankfurt: Verlag Peter Lang.

Gambetta, D. (1993). *The Sicilian mafia: The business of private protection.* Cambridge: Harvard University Press.

Gould, R. V. (2003). *Collision of wills: How ambiguity about social rank breeds conflict.* Chicago: University of Chicago Press.

Haller, M. H. (1991). *Life under Bruno: The economics of an organized crime family.* Pennsylvania: Pennsylvania Crime Commission.

Hess, H. (1998). *Mafia and mafiosi: Origin, power and myth.* New York: New York University Press.

Ianni, F. J. (1972). *A family business.* New York: Russell Sage Foundation.

Jamieson, A. (2000). *The antimafia: Italy's fight against organized crime.* New York: St. Martin's Press.

Klein, M. W., & Maxson, C. L. (1989). Street gang violence. In N. Weiner's (Ed.) *Violent crimes, violent criminals* (pp. 198–234). Beverly Hills: Sage.

Levitt, S. D., & Venkatesh, S. A. (2000). An economic analysis of a drug-selling gang's finances. *The Quarterly Journal of Economics, 115,* 755–789.

Loftin, C. (1984). Assaultive violence as contagious process. *Bulletin of the New York Academy of Medicine, 62,* 550–555.

Naylor, R. T. (1997). Mafias, myths, and markets: On the theory and practice of enterprise crime. *Transnational Criminal Organizations, 3,* 1–45.

Quinn, J., & Koch, D. S. (2003). The nature of criminality within one-percent motorcycle clubs. *Deviant Behavior, 24,* 281–305.

Reuter, P. (1983). *Disorganized crime: The economics of the invisible hand.* Cambridge: MIT Press.

Schelling, T. C. (1984). *Choice and consequence.* Cambridge: Harvard University Press.

Senechal de la Roche, R. (2001). Why is collective violence collective. *Sociological Theory, 19,* 126–144.

Sullivan, M. (1989). *Getting paid: Youth crime and work in the inner city.* Ithaca: Cornell University Press.

Thrasher, F. (1927). *The gang.* Chicago: Chicago University Press.

Tremblay, P., & Paré, P.-P. (2003). Crime and destiny: Patterns in serious offenders' mortality rates. *Canadian Journal of Criminology and Criminal Justice, 45,* 299–326.

Tremblay, P., Laisne, S., Cordeau, G., MacLean, B., & Shewshuck, A. (1989). Carrières criminelles collectives: évolution d'une population délinquante (les groupes motards). *Criminologie, 22,* 65–94.

Varese, F. (2001). *The Russian mafia: Private protection in a new market economy.* New York: Oxford University Press.

Wolf, D. R. (1991). *The rebels: A brotherhood of outlaw bikers.* Toronto: University of Toronto Press.

Yablonsky, L. (1959). The delinquent gang as a near-group. *Social Problems, 7,* 108–117.

Chapter 12
Controlling Human Smuggling in the Netherlands: How the Smuggling of Human Beings Was Transformed into a Serious Criminal Offence

Richard Staring

Just before the summer holidays of 2005, the Dutch National Criminal Investigation Service arrested ten members of a violent Chinese gang. All detainees were suspected of human smuggling, drug trafficking, extortion and liquidations. Police not only found large amounts of drugs and money, but also came across 24 automatic firearms, among which several Uzi's. The Chinese gang members were allegedly heavily involved in smuggling Chinese from their home country through the Netherlands to the United Kingdom. Smuggled immigrants had to pay up to forty or even fifty thousand euros for their journey and according to the police the gang members were prepared to use violence towards defaulters. In addition, the police suspected some of the arrested gang members of being linked to previously apprehended snakeheads of other Chinese smuggling organizations (Meeus, 2005). Since the turn of the millennium, the Chinese in the Netherlands have acquired the reputation of being involved in professionally organized human smuggling through Eastern European countries to the UK, using safe houses in harbour cities such as Rotterdam. The involvement of Chinese immigrants with human smuggling in the Netherlands became explicitly visible through the "Dover case." In this tragedy on June 18, 2000, 58 Chinese smuggled immigrants on their way to the United Kingdom suffocated in a refrigerated cargo container. The "Dover case" put the Netherlands on the map as a transit country for human smuggling although it wrongly focused solely on Chinese smugglers. It turned out that Dutch citizens from different ethnic backgrounds and with different judicial status were involved in the lucrative business of human smuggling (Staring, Engbersen, & Moerland, et al., 2005).

Until 1993, when human smuggling was included in penal law not much attention was given to human smuggling. Since then, politicians and policy makers have increasingly blamed human smugglers for sabotaging immigration and integration policies as well as for the influx of asylum seekers. More recently human smuggling is explicitly linked with the threat of international terrorism (see e.g. DNRI, 2005; IAM, 2000, 2002; IND, 1998). In an effort to combat human smuggling, special analysis and investigation services have been established that are solely responsible for combating human smuggling. In 2005, an amendment to the law on human smuggling was sent to the House of Representatives (TK 29291/1). One of the principal changes was the removal of the "profit element" from article 197a of the Criminal Code, which is in contrast to the UN Protocol against human

D. Siegel and H. Nelen (eds.), *Organized Crime: Culture, Markets and Policies.*
© Springer 2008

smuggling.[1] As a consequence, Public Prosecutors no longer have to prove that smugglers are motivated by profit. Instead, people accused of smuggling have to prove that they are altruistically or idealistically motivated. In 2004, the newly established National Criminal Intelligence department fighting organized crime in the Netherlands designated human smuggling as one of the seven main areas of organized crime needing their attention.

Human smuggling is also dealt with in the policy field of Dutch immigration control as is the case in most industrialized, labour-importing countries (Cornelius, 2004). Industrial countries increasingly try to control immigration by further tightening entry restrictions as well as by building up high administrative walls that marginalize and exclude illegal immigrants[2] from full participation in society. Simultaneously, these same countries have to deal with internal economic interests and push pressures in the sending countries as well as the structural demand for formal and informal labour in Western receiving countries. Besides, West-European governments are increasingly sensitive to public concerns about the consequences of the influx and settlement of immigrants for the identity and unity of the nation state.

The question is how the political actors, the media, and the public in the Netherlands have discussed and dealt with human smuggling in the context of these global developments. How did the political and public perception of human smuggling in the Netherlands develop over the last decades, and how has human smuggling become criminalized and adopted in criminal law and eventually become a priority in the Dutch fight against transnational organized crime? Another question posed in this article is to what extent these processes mirror the reality of the world of unwanted and unsolicited migration and human smuggling to the Netherlands. Does the volume and seriousness of organized human smuggling to and through the Netherlands justify the political and judicial attention it attracts? Before answering these questions, I will present a short historical description of the different law enforcement agencies and organizations in the Netherlands dealing with controlling human smuggling and human trafficking.

Dutch Law Enforcement Agencies Organizing the Fight Against Human Smuggling

All specific actors involved in the fight against human smuggling were established only after the introduction of criminal penalties on human smuggling at the end of 1993. The "fight against human smuggling" set off slowly and accelerated around

---

[1] The smuggling of migrants is defined in the United Nations Protocol Against the Smuggling of Migrants by Land, Sea and Air under article 3 as "the procurement, in order to obtain, directly or indirectly, *a financial or other material benefit*, of the illegal entry of a person into a State Party of which the person is not a national or a permanent resident" (UN protocol 2000: 3; italics added by author).

[2] Illegal immigrants are defined here as immigrants who do not have permission to enter or reside in the Netherlands. This definition includes people who overstay their visa; immigrants who come to the Netherlands without valid travel documents and have no valid residence permits; and asylum seekers who were denied entrance to the Netherlands, ignored the order to leave the country and stayed on without documents.

1997, when the Ministry of Justice started with a national approach and a special focus on human smuggling. Among others, the Taskforce Human Smuggling, successor to the unsuccessful Coordinating Consultation on Alien Smuggling (AOM), was initiated, but it only became effective in 1999 (Faber, 2002). The main tasks of this taskforce lie in controlling irregular migration and human smuggling. According to Faber (2002: 11) it is unclear exactly which organizations or institutions belong to this specific taskforce under the heading of the Department of Justice. Certainly one of them is the Information and Analysis Centre on Human Smuggling (IAM), established in 1999, in which the Royal Military Police (KMar), the Dutch Immigration and Naturalization Service (IND), as well as the National Criminal Intelligence Department (DCRI) collaborate. In the same year the Alien Smuggling Unit (UMS), a special investigative unit as part of the core team North and East Netherlands, also started its activities (TK 26345/1: 20–21).

Between 1993 and 1995, several so-called *core teams* were set up to combat supra-regional organized crime. Until the summer of 2003, members of these core teams were extracted from regional police forces. Only recently (January 1, 2004) these core teams were transformed into an independent National Criminal Investigation Service (Nationale Recherche). The core teams were formed around *geographical areas,* such as South America or Eastern Europe; around *infrastructures,* such as Schiphol Airport or the main port of Rotterdam; or around certain *types of organized crime,* such as human smuggling or synthetic drugs (van de Bunt, 2004). The core team that has a special Alien Smuggling Unit (UMS) resides at the core team "North and East Netherlands." After it became operational in 1997, this unit directed its attention fully on human smuggling and has since investigated numerous cases of organized human smuggling. In the same year, a national Public Prosecutor on Human Smuggling was appointed, who was also made responsible for prosecuting offences ranging from human trafficking to child pornography. In addition to this operational Alien Smuggling Unit and the employment of a specialized public Prosecutor in this field, 1997 witnessed the birth of the Information and Analysis Centre on Human Smuggling (*IAM*). The main goal of the centre is to gather (and update) information on the nature and size of human smuggling organizations related to the Netherlands. This means that the IAM not only analyses human smuggling within the irregular migration process, but also plays a central role in coordinating information (all data should be gathered by the IAM) as well as in proposing new criminal investigations in the field of human smuggling (Feenstra, 1999: 17–18). Next to the core teams and the UMS, the Royal Military Police (*Kmar*) investigates human smugglers and human smuggling organizations as part of their border enforcement duties.

In general, the fight against organized human smuggling is conducted at the level of core teams or specific units such as the UMS, often in collaboration with other services such as the IND, the IAM or the Royal Military Police. The Dutch Immigration and Naturalization Service (*IND*) is responsible for the enforcement of the Aliens Act. Within the context of human smuggling investigations, the IND is first of all responsible for providing relevant information to core teams on request. Like the IAM, employees of the IND analyze, for instance, asylum seekers' stories in order to describe and publish on human smuggling routes and trends.

Occasionally, these data are used for scientific analysis of human smuggling by independent researchers (see for instance Hesseling & Taselaar, 2001). The number of human smuggling cases that were brought before the Public Prosecution Service (PPS) slowly increased from 4 in 1994 to 268 cases in 2000. In the third millennium the number of cases decreased again to 130 cases in 2003. These figures tell us little about the supposed growth or decline in the activities of human smugglers. One could state, however, that the pattern in human smuggling investigations during the years mirrors the priority given to this type of crime by the PPS and the police, as well as the impact of the Dover transport disaster of 2000 (IAM, 2002: 94). The problems with "facts and figures" around human smuggling and trafficking are huge. According to Salt and Hogarth (2000: 31) there is, first of all, a lack of data, and if available it is often unclear how the data were gathered. According to these IOM researchers most statistical data on the number of smuggled people are at best rough estimates based on a set of assumptions. Furthermore, in most countries there is no central registration of human smuggling and/or trafficking; different institutions mean different definitions and ultimately poor and incomparable figures. Thirdly, human smugglers are not always prosecuted on 197a of the Dutch Criminal Code but in many cases on different grounds as, for instance, organized crime, forgery or false documents. Notwithstanding these limitations on the figures around human smuggling, it is certainly perceived by governmental actors as one of the top priorities in the global fight against transnational organized crime.

Controlling Immigration in the Netherlands

One can discern several global developments as well as local incidents that are crucial to understanding the debate on human smuggling in the Netherlands and the way it was implemented in Dutch law. Globalization is nothing more than a process in which "people and places in the world are becoming more extensively and densely connected to each other as a consequence of increasing transnational flows of capital/goods, information/ideas, and people" (Kalb, 2000: 1). Like all western industrialized countries, the Netherlands was confronted, along with other transnational flows, with this movement of people, among which illegal immigrants and asylum seekers (cf. Staring, 2000). Processes of international migration have many facets, but one important observation in the Dutch context relates to a gradual process of criminalization of illegal immigrants and asylum seekers during the last decades which evolved, from the turn of the millennium onwards, into a rather intolerant political and public climate towards the undocumented and asylum seekers.

According to Muus (2004: 265), the immigration control policies of the Netherlands can be characterized as rather pragmatic, as "an inevitable outcome of economic needs and humanitarian considerations." He describes Dutch immigration policies during the last decades as largely ad-hoc, reactive instead of "proactively shaping the conditions for immigration" (ibid. 266). In retrospect, the overall tendency is towards a more restrictive Dutch migration policy, a discouragement of illegal stay, and a further tightening of the opportunities for illegal aliens on the formal labour market and in gaining access to the facilities of the welfare state. The focus of immigration policies slowly shifted from external to internal control; from controlling physical borders to raising barricades through administrative measures (cf. Entzinger, 2002). Contrary to what policy-makers initially expected, most of the immigrants stayed and initiated further migration. The absolute number of immigrants – or ethnic minorities – gradually grew from 206,000 in 1971 to 648,000 in 1985 to around 1,720,000 people with a non-western background living in the Netherlands in 2006 (Statistics Netherlands, 2006). As a consequence more then 10% of the total population residing in the Netherlands by 2006 has a non-western background (Statistics Netherlands, 2006).[3] To a large extent these immigrants live in urban areas.

During the last four decades we have witnessed a social reclassification of illegal immigrants from "spontaneous labourers" to "undesired and excluded illegal immigrants" (Engbersen & van der Leun, 2001). This reconfiguration of the "spontaneous labour migrant" into the "new untouchable" coincides with increasingly tougher control on illegal stay and entry.[4] For the moment, the formal exclusion of illegal immigrants has culminated in two complementary white papers *Return policy. Measures for a more effective enforcement of return policies* (TK 29344/1, 2003) and *Illegal Aliens* (DVB, 2004). Following measures aimed at reducing the asylum influx, these white papers stress the necessity of an effective return policy combined with measures aiming at discouraging illegal stay. In the document *Illegal aliens*, Minister Verdonk of Immigration and Integration proposed several correlated measures to complicate and discourage the illegal stay of aliens. Among the central domains of action are more severe measures against those people who profit from the illegal stay of aliens. Exploiters such as shady landlords, unreliable employers and especially human traffickers, as they "not only profit but sometimes also violate human rights" (2004: 5), should be dealt with more efficiently. According to the Minister of

---

[3] The figure of almost 1,720,000 persons with a non-Western background in this text refers to first and second-generation immigrants with a non-Western background. The total number of people with a foreign background is over 2,900,000 (Netherlands Statistics/Statline 2006). According to the broad Netherlands Statistics definition, people have a foreign background if at least one parent was born abroad. Netherlands Statistics also differentiates between people born abroad (first generation) and people born in the Netherlands (second generation). Foreign background is determined by the country of birth of the person (first generation) or the country of birth of the person's mother (second generation). If the mother was born in the Netherlands, the person is classified according to the father's country of birth. People with a non-Western background come from Turkey and countries in Africa, South America and Asia except for Indonesia and Japan. Indonesia and Japan are categorized with the Western countries on the basis of their socio-economic and socio-cultural position. For more detailed information see http://statline.Netherlands Statistics.nl.

[4] Among the most important laws that were implemented during the 1990s that had a direct impact on the position of illegal immigrants in Dutch society are the linking of the social-fiscal number to a valid residence status in 1991, which excluded illegal immigrants from participating in the formal labour market, the Compulsory Identification Act of 1994, the Marriage of Convenience Act of 1994, the Linking Act in 1998 and the Aliens Act in 2000.

Immigration and Integration, structures that facilitate human trafficking encourage illegal stay. By tackling these criminal structures an important contribution to the approach of illegal stay will be provided (ibid. 14).[5] According to Cornelius (2004), most attempts by the Dutch government to narrow the gaps between immigration control policies and its outcomes, especially in the field of irregular migration, have not seriously altered the image of the Netherlands as a welcoming country and they also come with unintended consequences. Especially in the field of entry and irregular migration these policy efforts seem to have resulted in a stronger position of human smuggling organizations (2004: 30).

From Tolerated Immigrants to Terrorists

During the 1980s, when policy makers realized that the former guest workers were going to stay in the Netherlands, a group-based integration policy was developed, which included the "maintenance of the immigrants' own identity." In the 1990s, however, this policy was increasingly criticized for its lack of success and it gradually changed into a more individually based integration concept (Ghorashi, 2004: 164). Bolkestein, the then leader of the conservative liberal party (VVD), was the first to emphasize the negative (economical, social, and cultural) effects of immigration on Dutch society. Bolkestein and his successors defended a tough integration approach towards immigrants which came close to advocating assimilation, rather than leaving room for the preservation of immigrant cultures or identities (ibid.).

More in general, the public attitude towards immigrants in Dutch society, particularly immigrants with a Muslim background, hardened during the 1990s, especially around the turn of the millennium. Politicians, academics, as well as the public began to question the success of the Dutch integration policy and the concept of the Netherlands as a multicultural society. Especially "traditional" Muslim immigrants, more in particular the sizable Turkish and Moroccan communities, were blamed for their lack of integration into Dutch society, their poor position on the labour market, high unemployment rates and involvement in criminal activities (the latter was mainly directed at young Antilleans and Moroccan Dutch).

A number of events and violent incidents are important to understanding this farewell to "Dutch tolerance." First of all, there was the publication in a leading newspaper of an essay by publicist and journalist Scheffer (2000) on the failure of Dutch integration policies entitled "A multicultural drama," which had a significant impact on the national debate. Then there were the events of September 11, 2001. Not much later, the Netherlands witnessed the political success of former sociology professor Pim Fortuyn and his national party *Lijst Pim Fortuyn*, followed by his assassination nine days before the parliamentary elections of May 15, 2003. In 2004, Theo van Gogh, publicist and director of controversial movies including *Submission*, which

---

[5] It appears that the term "human smuggler" is confused with "human trafficker."

takes a critical look at Islamic culture, was murdered by an Islamic extremist. All these events have contributed to a polarized Dutch society, where many "native" Dutch citizens are distrustful of Muslim immigrants. The international media also focussed on the killing of Van Gogh. *Migration News*, a reputable journal on global migration and related issues reported the news as follows:

> Dutch filmmaker Theo van Gogh, who mocked conservative Muslims, was killed in November 2004 by a man the police described as a Muslim extremist, setting off a debate about whether Europe's most liberal society could be fragmenting, with the violence a harbinger of integration debates in other European countries. The government's response has been to promise more money to fight terrorism and stricter immigration laws (*Migration News* 5-1-2005).

It is exactly this explicit linking of anti-terrorism measures with immigration policies that led the European Monitoring Centre on Racism and Xenophobia to point at the growing atmosphere of insecurity and intolerance in the Netherlands, especially where Muslims are presented as an "internal security threat" (MN 04/01/2002). It is exactly this "perpetual sense of crisis" that, according to Garland (2001), strongly contributes to a culture of control; something also observed by Cornelius (2004) with respect to immigration.

In the context of the incidents described and given the way many politicians reacted, the idea that especially Muslim immigrants in the Netherlands should integrate better and faster could flourish. The Dutch government emphasized that successful integration could only be accomplished if international migration flows to the Netherlands could be restricted and controlled. Political statements such as these paved the way for measures favouring immigration control, including laws designed to improve border control and to prevent human smuggling. Other elements that contributed to these restrictive measures can be found in the asylum debate.

Human Smuggling and Asylum Seekers

Various government documents stress the importance and responsibility of human smugglers for the nature and size of migration flows, especially for the arrival of asylum seekers in the Netherlands (see e.g. DNRI, 2005; IAM, 2000, 2002). In 1999, the spokesman for the Public Prosecution Service stated that human smugglers facilitated the entrance of two thirds of all asylum seekers into the Netherlands. According to the spokesman, human smugglers routinely used centres for asylum seekers for their activities (ANP July 13, 1999).

The increasing influx of asylum seekers during the last decade fuelled one of the dominant political and public debates during the 1990s and prompted the Dutch government to introduce increasingly restrictive asylum policies. Especially during the end of the 1990s the number of applicants rose to more than 40,000 over the year. Among policymakers there was a general belief that the Netherlands, due to its generous asylum procedures, was among the most attractive countries in the European Union. In 2001, this upward trend was reversed; the number of applicants fell below 10,000 in 2004.[6] Between 1997 and 2001, over 40% of the newly arrived immigrants came within the context of family reunification (21.3%) and marriage (19.5%). Almost 20% of the influx during this period could be labelled as labour migration (19.5%), while approximately 13% came to study or for other reasons. Just slightly more than 25% of the newly arrived can be labelled as asylum seekers (26.3%) (Minderhoud, 2004: 13). This would suggest that the "asylum problem" (and the related topic of human smuggling) figured too prominently in the migration policy discussions.

The arrival of asylum seekers gave rise to several debates, among which the portrayal of asylum seekers as economic refugees and/or criminals stands out. The establishment of new reception centres invariably evoked protests from local residents who were worried about the price of their real estate and afraid of increasing insecurity and rising crime figures. The centres were perceived as "universities for crime" or as "hotbeds for organized crime such as human trafficking and drugs," and these messages were widely covered by the national media (De Volkskrant, May 2, 1998). Human smuggling organizations supposedly used the asylum centres as cheap hotels for their clientele in transit to the United Kingdom or other countries. During the 1990s, the media increasingly associated and equated asylum seekers with economic refugees who, in their search for economic benefit and in cooperation with human smugglers, exploited the Dutch reception centres for their own gain. According to these media reports, the smugglers used the reception centres as "safe houses" before smuggling the "asylum seekers" to the United Kingdom or the Scandinavian countries (see for instance De Volkskrant, July 14, 1999).

Organized Crime and Terrorism in the Netherlands

A second manifestation of globalization and migration has to do with transnational organized crime and international terrorism (Bunt, 2004). According to Paoli and Fijnaut (2004: 3), globalization processes have "accelerated the interconnection between previously separate domestic illegal markets and increased the mobility of criminals across national borders." Especially during the 1990s, the Netherlands saw a growing political, judicial, as well as academic interest in transnational organized crime, including human smuggling. Finally, after September 11, human smuggling was increasingly linked with international terrorism. These processes have contributed to a negative approach to all different modes of gaining entrance to the Netherlands, be it legally by way of marriage to a partner from abroad, or clandestine with the assistance of human smugglers.[7]

---

[6] In 2005, however, the number of asylum requests slightly increased to 12.000 (http://www.statline.nl).

[7] There is, for instance, a proposal to reduce the number of immigrants who settle in the Netherlands through marriage by restricting some of the conditions for marrying a spouse from abroad.

The 1990s can be characterized by a growing concern for organized crime in the Netherlands. According to criminologist Bunt (2004: 3), "the general public as well as the politicians [were] seriously concerned about the problem of organized crime and both were in favour of far-reaching measures." At first, the attention was mainly concentrated on drugs; it took some time before human trafficking and smuggling were also included in the focus on organized crime. Around 1990, several studies by scientists, police analysts and journalists were published. They all presented a picture of Dutch organized crime that convinced the public and the politicians that the Netherlands were faced with a serious problem which had been underestimated until then (Middelburg, 1988; van Duyne, Kouwenberg, & Romeijn, 1990; cf. van de Bunt, 2004: 6–7). In 1992, a white paper on organized crime was published advocating both a repressive as well as a preventive approach to combat organized crime. The safety of the general public was explicitly mentioned in the paper, which stands in contrast to one of the latest white papers on crime *Towards a safer society* (Ministerie van Justitie, 2002) in which "security" is at the centre of interest.[8] Apparently, organized crime plays no part in the discussion around security and measures aimed at increasing security, as there is no mention of organized crime in the document.

A scandal over the "controlled delivery" of drugs into the Netherlands by police informants under the guidance of interregional investigation teams in 1993 resulted in a "parliamentary inquiry into the methods used by the police and the Public Prosecutor's Service to investigate organized crime" starting in October 1994 (van de Bunt, 2004). The final report (Parlementaire Enquetecommissie Opsporingsmethoden, 1996) by the committee, partially written by a team of criminologists, presented an picture of the various forms of organized crime in the Netherlands and the way it had become embedded in Dutch society. The commission also criticized the various judicial bodies and organizations engaged in fighting organized crime for not exchanging enough information. In the final report, much attention was paid to organized crime being involved in the drug business. Human smuggling and human trafficking were dealt with only marginally.

The policies developed over the last decade are both repressive, aimed at "catching criminals," as well as preventive, i.e. focusing on the various circumstances facilitating organized crime instead of focusing solely on the perpetrators. The preventive approach entails, amongst other measures, efforts to tackle money laundering and to improve the integrity of public administration.[9] As part of this preventive approach, local governments, citizens, and private enterprises are increasingly held accountable for reducing the opportunities for organized crime. Van Swaaningen (2004) observed that this general trend of responsibilisation of

---

[8] Security, integration, health care, and education were listed as priorities by the coalition government formed in May 2003 by the Christian Democrats (CDA), the Liberal Party (VVD) and the List Pim Fortuyn (LPF) (Government Policy Statement, Balkenende I, 2003).

[9] The MOT-Act (Act on the Disclosure of Unusual Transactions), introduced in 1994, made it mandatory for financial institutions to report unusual financial transactions.

non-penal actors is less influential in the Dutch context than it is in the United States or in the UK. Nevertheless, with respect to human smuggling and illegal immigration into the Netherlands, passenger transport companies are now responsible for carefully examining the travel documents of their passengers at the risk of being fined or otherwise penalized (van de Bunt, 2004: 20; cf. Pieters, 2006).

Organized crime is often linked with concerns of corruption and collusion by the licit world. Detailed studies on organized human smuggling and trafficking conclude that the activities of smugglers are generally not dependent on infiltration or corruption (Staring et al., 2005). Human smugglers usually rely on existing transportation opportunities, without the need for involving carrier employees or border control officers.[10] This means that measures aimed at improving the integrity of (public) administration will have little impact on the prevention of organized human smuggling.

Transnational organized crime in the Netherlands is usually depicted as an internal threat to society. As a result, attention was initially focused on researching already present or incoming forms of organized crime. Much less attention was paid to "transit oriented organized crime" or "export oriented organized crime." This preference for "incoming crime" was also mirrored in the preferences of the agencies involved in the fight against organized crime. The majority of police investigations into organized crime focused around the illegal drug business (Bunt, 2004: 33). Other lucrative forms of (transnational) organized crime, such as human smuggling, lacked priority. It took some time before the Dutch government called for more attention to criminal networks active in this specific field (Ministerie van Justitie, 2001: 33). When the National Criminal Investigation Unit was established in 2002, one of its goals was to acquire knowledge on and to intervene in the logistics of organized crime, with explicit reference to the transit character of organized crime in the Netherlands. As mentioned earlier, human smuggling is included among its seven main areas of attention.

At the end of the 1990s, political and public interest in organized crime slowly decreased. According to Van de Bunt, one of the reasons for this lack of interest can be found in the fact that other crimes such as street violence and the issue of "insecurity" demanded more attention. The factual knowledge gathered by scientists on the subject of organized crime was another important factor. The "moral panic" about organized crime of the early 1990s gave way to a more realistic assessment of the threat posed by this phenomenon based on facts produced by criminologists and other social scientists (Bunt, 2004: 13). In the meantime, a highly developed infrastructure to combat organized crime has been developed in the Netherlands.

---

[10] When corruption is mentioned in the context of human smuggling it is mostly attributed to officials in the immigrants' source countries. There have been several incidents, however, (especially at Schiphol Airport) where employees were involved in human smuggling. In 2005, the Kmar published figures of the number of employees of Schiphol Airport arrested for human smuggling. During the last five years, 675 Schiphol employees - among which some members of the Kmar themselves - were apprehended for assisting illegal immigrants entering the Netherlands (*NRC Handelsblad*, January 20, 2005).

Law enforcement agencies have identified both human smuggling and human trafficking as priority targets. Although there is an on-going interest from the media and from the public in the increasing number of assassinations of "well-known Dutch criminals,"[11] attention seems to have shifted to the new "face of insecurity," i.e. Muslim fundamentalism and Islamic terrorism.

Within two months of the terrorist attacks by Osama Bin Laden's Al Qaeda on US targets on September 11, 2001, the Dutch government formulated a *Framework for action on terrorism control and security*. By June 2003, six "progress reports" had been published on how the Dutch government planned to cope with the threat of international terrorism. The action programs focused on the prevention of terrorism, the protection of the public or social order, and the investigation and prosecution of terrorist crimes. Some of the main points of action dealing with (or at least affect) human smuggling and irregular migration movements, as for instance, the increase of control at the Dutch external borders (airports and harbours), and the exploration of the possibilities of biometrical identification. Besides these more indirect measures, one of the action points is targeted directly on controlling human smuggling. Several measures are proposed, summarized by an enlargement of the broadly composed capacity of the special task force Unit Human Smuggling (UMS). Three elements are put forward in this first *Framework for action on terrorism control and security*. First, to improve the information position on the interrelatedness of human smuggling and terrorism by, secondly, increasing the capacity to analyze this information, and thirdly, by increasing the opportunities to exchange this information between all organizations involved (2001: 12). In the following frameworks for action these thoughts are operationalized by increasing the capacity of the Unit Human Smuggling, the IAM, the Public Prosecution Service, the taskforces of Transnational Crime (GOC), as well as the Royal Military Police.

In their second report on human smuggling, the IAM (2002) concluded that so-called "Islamic terrorism" is closely connected to illegal immigration and as such the IAM points to a "security risk." From the IAM's (2002: 131) perspective, one should seriously take into account the fact that terrorist groups employ false documents in order to achieve their goals. Although members of these terrorist networks do not utilize commercial human smugglers for their transport between different countries, they do rely on the practice of smuggling to carry out their attacks (ibid.: 136–138). In their latest "state of the art," the IAM cautiously concludes that terrorists, in order to finance their activities, rely, among others, on the profits made from human smuggling (DNRI, 2005). Clear evidence or convincing arguments that could underscore this link between terrorism and human smuggling can, however, not be presented. The central point is that by focusing on terrorism, human smuggling is not only dealt with more intensively in terms of control, but human smuggling is also increasingly connected to and associated with the fear of Islam and Islamic fundamentalism.

---

[11] For an overview of these assassinations until May 2004, see *De Volkskrant*, May 18, 2004: 11.

But what empirical evidence is there to support these restrictive policies and judicial priorities? Through different processes human smuggling is increasingly criminalized and perceived as a threat to society in general and to successful integration of its minorities. An enormous apparatus has been developed to fight human smuggling, but again, are these developments embedded in the realities of immigrant lives? How serious is human smuggling in or through the Netherlands? In order to shed some light on these questions some empirical research findings on irregular migration patterns will be discussed below.

How do Illegal Immigrants and Asylum Seekers Enter the Netherlands?

Research among illegal immigrants illustrate that the majority of the illegal immigrants do not come into the Netherlands through the help of criminal organizations but enter legally. Elsewhere I described extensively how illegal immigrants enter the Netherlands by utilizing different sources of social capital (Staring, 2004). Out of an analysis of over three hundred stories of illegal immigrants living in the Netherlands it turned out that two thirds of the undocumented immigrants arrived with no support of (commercial) human smugglers. The majority (around 55%) came by way of a visa (especially tourist visa) supported by relatives or friends or were smuggled in by family or friends, living in the Netherlands. Another 10% of the illegal immigrants came on their own without the help of kin, friends or smugglers. The same analysis views that around one third of the illegal immigrants – especially those who applied for asylum in the Netherlands – were facilitated by human smugglers one way or another (Staring, 2004; cf. Kleemans & Brienen, 2001).

In the first and by far most dominant migration pattern, the arrival of illegal immigrants is guided by the managing efforts of supportive and loyal relatives in transnational networks. Most illegal immigrants are embedded to some degree in transnational networks. These geographically scattered networks, especially kinship-based links, explain the arrival and further incorporation of illegal immigrants in the Netherlands. For these socially embedded illegal immigrants there was no need for the services of human smugglers as more convenient opportunities were available to them. The comprehensive support provided in this migration pattern is mainly provided by their relatives, or, to a lesser extent, by their friends or acquaintances. Illegal immigrants normally know where they are going. Relatives abroad financially support them in their trip or enable them to obtain the travel documents they need. After the newcomers arrive, the transnational households welcome them and help them to settle in and find work. These patterns coincide with what is referred to in the literature as chain migration, although in an irregular form (cf. MacDonald & MacDonald, 1974; Price, 1963). In the second migration pattern, prospective immigrants largely travel on their own with no valid travel documents. Illegal immigrants from Eastern Europe, especially from the former Soviet Union and Yugoslavia, follow this migration pattern. Partly due to their geographical proximity, they can travel relatively easily and by their own means of transport to the Netherlands. They are often highly educated, mingle without too much difficulty with the native Dutch population, and consequently have extensive contact with them. They frequently have a Dutch employer or find a formal or informal job through a native Dutch mediator. Their own ethnic group is not important to these illegal immigrants, who often speak Dutch or English fluently. They mainly move in the Dutch world and aspire to settle permanently in the Netherlands, especially by legalizing their stay by marrying a Dutch person.

Both migration patterns that fuel unsolicited migration flows, go along without the involvement and support of human smugglers and are responsible for the arrival of two thirds of the illegal immigrants. These findings are not unique; the legal embeddedness of much of the irregular migration flows has been reported by other researchers in the Netherlands as well as for other European countries (see for instance Liempt & Doomernik, 2006; Spener, 2004). As such, these supportive networks are far removed from the criminal organized human smuggling networks that are at fore front of and became one of the top priorities of Dutch and other western societies in their fight against (organized) crime.

In addition to the empirical research among illegal immigrants, we analysed eleven large police investigations of organized human smuggling, all situated around the harbour city of Rotterdam (Staring et al., 2005). Within these police investigations, thirteen human smuggling networks could be distinguished. In this report we developed a continuum of human smuggling organizations in which we differentiate between the so-called "negotiating criminal network" and at the other end of the continuum the so-called "commanding criminal network." The human smuggling networks constructed out of the police investigations are classified according to two important network characteristics we borrowed from the organiza-tion sociology. *Domination*, the first dimension, was analyzed by looking at who is in control and how control is empowered within the criminal network and also by looking at the kind of relationship between gang members and their customers. Violence or the threat of violence within the smuggling network as well as towards the smuggled immigrants was also taken in to account. *Coordination*, the second network dimension, was explored by distinguishing the different tasks within the process of human smuggling and by looking at who takes care of these tasks. Besides these two dimensions we also looked at the transnational embeddedness of the criminal networks.

The "commanding criminal network" resembles the classic image of a hierarchic criminal organization with a clear leader at the top who is in control of his subordi-nates. Money and (the threat of) violence are important mechanisms to keep control over the gang members. The relationships with their customers can be characterized at best as business-like although appointments between the two parties were difficult to make by the smuggled immigrants. Abduction as well as mistreatment of smuggled immigrants by gang members in order to receive higher fees were commonplace. In terms of coordination these criminal networks can be characterized by its large scale and by the many people involved often performing only a limited numbers of tasks. These criminal networks were in terms of ethnic composition more heterogeneous than the "negotiating criminal network." As with the "negotiating criminal network," this "commanding criminal network" is strongly embedded within the communities of fellow countrymen, although outsiders hardly perform tasks within the human smuggling network.

Within the "negotiating criminal network" it is much more difficult to envision a specific leader or a hierarchic structure. No one really seems to be in control of another member of the criminal network and (the threat of) violence is absent between the smugglers as well as towards the smuggled immigrants who are free to go as they like. The relationships within the criminal networks, as well as between the smugglers and their customers, are much less business-like and much more dominated by kinship, friendship or ethnicity. Prices smuggled immigrants have to pay are much closer towards the "real costs" than the exorbitant prices that circulate within the "commanding criminal networks." Negotiating criminal networks' are much more small-scale networks where few members perform many tasks. Simultaneously, there are many outsiders who once in a while engage in the smuggling activities. These "negotiating criminal networks" are also characterized by their ethnic closure: smugglers as well as their customers belong to the same ethnic group.

Dutch police classified all human smuggling organizations we studied under the heading of organized crime. The cases of the Chinese smuggling organizations that dominate the media attention fit perfectly well on the image of the "commanding criminal network," but the majority of the networks analyzed in *The social organization of human smuggling*, however, resembled much more the "negotiating criminal network." In a way one could state that these human smuggling organizations answered much more to the image of small-scale informal entrepreneurship where economic motives intermingle with kinship based loyalties. Large-scale, violent, and professionally organized human smuggling networks, such as the Chinese gangs, still seem to be an exception. From the perspective of the risks for the people being smuggled, this also implies that the majority of the smuggled immigrants travel legally with the required travel documents, with relatives, on their or own or in rather small groups. Only a minority of the smuggled immigrants come within large groups with all the physical risks involved that go along with these modes of travelling.

Concluding Remarks

During the last decade, human smuggling became associated with the world of transnational organized crime and international migration control. In this contribution, several global developments and incidents within the Netherlands have been described that have contributed to a climate in which human smuggling has evolved from a non-problematic event to a serious criminal offence threatening to derail Dutch immigration and integration policies and ultimately threatening national security by facilitating international terrorism. Human smuggling in the Netherlands was, above all, prioritized from the top, starting at first at the level of the European Union's fight against organized crime and the disappearance of physical borders.

Human smuggling is thus not born as a "priority from below"; as an urgent response to a need severely felt by different actors on the political and judicial level.

Among these processes of globalization, the focus in this contribution was on the flows of people as labour migrants or as refugees, transnational organized crime and international terrorism. As in most other western countries, migration policies in the Netherlands are increasingly characterized in terms of migration control. For the Netherlands this meant a shift from reactive and pragmatic policies towards proactive policies. The Dutch government increasingly tries to control global migration flows by building up administrative walls. Simultaneously migration policies focus on the integration of the settled immigrants at the expense of marginalizing and excluding illegal immigrants. During this policy transformation, immigrants without valid residence papers were reclassified from spontaneous labour immigrants during the 1960s to illegal immigrants as undesired freeloaders or even criminals in the third millennium. This reclassification was fuelled by three processes: a changing attitude towards long-term unemployment in the Netherlands; the notion that the integration of settled immigrants can only be achieved by restricting the entrance of newcomers; and by the growing number of asylum seekers during the 1990s. In debates around the Dutch asylum procedure, asylum seekers were increasingly portrayed as economic fortune seekers and criminals benefiting from a generous Dutch procedure.

The political, judicial, as well as academic attention for transnational organized crime such as human smuggling and trafficking in human beings in the Netherlands during the 1990s resulted in a highly developed infrastructure of organizations and law enforcement agencies dealing with organized crime and later on with the threat of international terrorism. Directed by European Union measures as well as harrowing incidents as the Dover case, human smuggling became criminalized and was placed high on the political agenda in the Netherlands, but to some extent remained a paper tiger. Furthermore, the fragmented structure of the different agencies dealing with human smuggling, lead to poor cooperation and hindered effective investigations.

These global processes and local incidents created a feeling of fear and insecurity in the Netherlands in which more restrictive migration policies and changes in criminal law could be advocated and were implemented. These developments also paved the way for further amendments to the law on human smuggling. The empirical research done in the field of human smuggling does not mirror the threatening organized crime image that fuelled the judicial developments around human smuggling. As described, most of the unwanted immigrants enter the Netherlands through legal channels with the help of relatives or close friends living in Europe and the Netherlands. In most cases these unsolicited visitors travel on their own with the required documents, risking as much as a criminologist visiting a conference abroad. Even in those cases where immigrants pay smugglers for their entry into the Netherlands, these networks labelled as "organized crime" resemble much more the family-like migration patterns than the criminal smuggling gangs described in the introduction. This means that the judicial fight against human smuggling as a form of organized crime should be seriously reconsidered. Why further criminalize the migration process and simultaneously making entry into the European Union more difficult and an even more profitable market for criminal gangs.

References

Afrekeningen in Amsterdam en omstreken. (2004, 18 May). *De Volkskrant*, sectie Binnenland, p. 11.
Bunt, H. G. van de. (2004). Organised crime policies in the Netherlands. In C. Fijnaut & L. Paoli
    (Eds.), *Organised crime in Europe. Concepts, patterns and control policies in the European
    Union and beyond* (pp. 677–716). Dordrecht: Springer.
Cornelius, W. (2004). Controlling immigration: The limits of governmental intervention. In W. A.
    Cornelius, T. Tsuda, P. Martin, & J. Hollifield (Eds.), *Controlling immigration. A global per-
    spective* (pp. 3–48). Stanford: Stanford University Press.
DNRI. (2005). *Mensensmokkel in beeld, 2002–2003*. Zoetermeer: KLPD-DNRI.
Duyne, P., Kouwenberg, R. van, & Romeijn, G. (1990). *Misdaadondernemingen: ondernemende
    misdadigers in Nederland*. Deventer: Gouda Quint.
DVB. (2004). *Illegalennota. Aanvullende maatregelen voor het tegengaan van illegaliteit en de
    aanpak van uitbuiters van illegalen in Nederland*. 5282929/04, April 23, 2004.
Engbersen, G., & Leun, J. van der. (2001). The social construction of illegality and criminality.
    European Journal of Criminal Policy and Research, 9, 51–70.
Entzinger, H. (2002). *Voorbij de multiculturele samenleving*. Assen: Van Gorcum.
Faber, W. (2002). *De macht over het stuur. Onderzoek besturingssysteem aanpak mensensmokkel*.
    Oss: Faber Organisatievernieuwing.
Feenstra, C. (1999). Trends in mensensmokkel. In W. de Bruijn, H. de Jong, H. Pauwels, &
    I. Voorhoeve (Eds.), *Mensensmokkel naar Nederland. Een bundeling van lezingen over
    mensensmokkel in relatie tot de bestrijding van de (georganiseerde) criminaliteit* (pp. 9–20).
    Gravenhage: Elsevier.
Garland, D. (2001). *The culture of control. Crime and social order in contemporary society*.
    Oxford: Oxford University Press.
Ghorashi, H. (2004). Ayaan Hirschi Ali: Daring or dogmatic? Debates on multiculturalism and
    emancipation in the Netherlands. *Focaal – European Journal of Anthropology, 42*,163–169.
Hesseling, R., & Taselaar, A. (2001). Asielmigratie en mensensmokkel. *Tijdschrift voor
    Criminologie, 43*(4), 340–349.
IAM. (2000). *Dreigingsbeeld 2000. Inzicht in aard en omvang van mensensmokkel in relatie tot
    Nederland*. Zoetermeer: Informatie- en Analyse centrum Mensen-smokkel.
IAM. (2002). *Mensensmokkel in beeld 2000–2001*. Zeist: Uitgeverij Kerckebosch BV.
IND (Immigration and Naturalization Service). (1998). *Keten in kaart: trends en ontwikkelingen
    in de vreemdelingenketen 1998*. Den Haag: Ministerie van Jusitie, IND.
Kalb, D. (2000). Localizing flows: Power, paths, institutions, and networks. In D. Kalb et al.
    (Eds.), *The ends of globalization. Bringing society back in* (pp. 1–29). Lanham: Rowman &
    Littlefield.
Kleemans, E. R., & Brienen, M. (2001). Van vriendendienst tot slangenkop. *Tijdschrift voor
    Criminologie, 43*(4), 350–359.
Liempt, I. van, & Doomernik, J. (2006) Migrant's agency in the smuggling process. The perspec-
    tives of smuggled migrants in the Netherlands. *International Migration, 44*(4), 165–190.
MacDonald, J., & MacDonald, L. (1974). Chain migration, ethnic neighborhood formation, and
    social networks. In C. Tilly (Ed.), *An urban world* (pp. 226–235). Boston/Toronto: Little,
    Brown and Company.
Meeus, J. (2005). Politie pakt Chinese gangsters. *De Volkskrant*, 25 June 2005.
Mensensmokkelaars brengen tweederde deel asielzoekers binnen. (1999, July 13). *ANP,
    Binnenland* (Algemeen Nederlands Persbureau).
Middelburg, B. (1988). *De mafia in Amsterdam*. Amsterdam: De Arbeiderspers.
Migration News 5–1–2005.
Minderhoud, P. (2004). Het immigratiebeleid. In Han Entzinger en Jelle van der Meer (Eds.),
    Grenzeloze solidariteit: naar een migratiebestendige verzorgingstaat. Amsterdam: De Balie.
Ministerie van Justitie en Ministerie van Binnenlandse Zaken en Koninkrijksrelaties. (2001). *Nota
    criminaliteitsbeheersing. Investeren in een zichtbare overheid*. Den Haag: Ministerie van Justitie.

Ministerie van Justitie, Ministerie van Binnenlandse Zaken en Koninkrijksrelaties. (2002). *Naar een veiliger samenleving*, Den Haag, oktober 2002.: http://www.justitie.nl.

Muus, P. (2004). The Netherlands: A pragmatic approach to economic needs and humanitarian considerations. In W. A. Cornelius, T. Tsuda, P. Martin, & J. Hollifield (Eds.), *Controlling immigration. A global perspective* (pp. 263–288). Stanford: Stanford University Press.

NRC Handelsblad. (2005, January 20). Schiphol: 675 man vast wegens mensensmokkel.

Parlementaire Enquetecommissie Opsporingsmethoden. (1996). *Inzake opsporing*. Den Haag: SDU.

Pieters, B. (2006). Dutch criminal and administrative law concerning trafficking in and smuggling of human beings. In E. Guild & P. Minderhoud (Eds.), *Immigration and criminal law in the European Union. The legal measures and social consequences of criminal law in member states on trafficking and smuggling in human beings* (pp. 201–240). Leiden/Boston: Martinus Nijhoff Publishers.

Price, C. A. (1963). *Southern Europeans in Australia*. Melbourne: Oxford University Press.

Salt, J., & Hogarth, J. (2000). Part I. Migrant trafficking and human smuggling in Europe: A review of the evidence. *Migrant trafficking and human smuggling in Europe. A review of the evidence with case studies from Hungary, Poland and Ukraine*. Geneva: International Organization for Migration.

Scheffer, P. (2000). Het multiculturele drama. *NRC Handelsblad*, 29 januari 2000: p. 6.

Spener, D. (2004). Mexican migrant-smuggling: A cross-border cottage industry. *Journal of International Migration and Integration, 5*(3), 295–320.

Staring, R. (2000). Flows of people: Globalization, migration, and transnational communities. In D. Kalb et al. (Eds.), *The ends of globalization. Bringing society back in* (pp. 203–215). Lanham: Rowman & Littlefield.

Staring, R. (2004). Facilitating the arrival of illegal immigrants in the Netherlands: Irregular chain migration versus smuggling chains. *Journal of International Migration and Integration, 5*(3), 273–294.

Staring, R. (2006). Controlling immigration and organized crime in the Netherlands. Dutch developments and debates on human smuggling and trafficking. In E. Guild & P. Minderhoud (Eds.), *Immigration and criminal law in the European Union. The legal measures and social consequences of criminal law in member states on trafficking and smuggling in human beings* (pp. 241–270). Leiden, Boston: Martinus Nijhoff Publishers.

Staring, R., Engbersen, G., Moerland, H., et al. (2005). *De sociale organisatie van mensensmokkel*. Zeist: Kerckebosch.

Statistics Netherlands 2006.

Swaaningen, R. Van. (2004). Veiligheid in Nederland en Europa. Een sociologische beschouwing aan de hand van David Garland. *Justitiële Verkenningen, 30*(4), 9–23.

Tweede Kamer, vergaderjaar 1998–1999, 26345, nr. 1.

Tweede Kamer, vergaderjaar 2003–2004, 29291, nr. 1, Uitvoering van internationale regelgeving ter bestrijding van mensensmokkel en mensenhandel.

Tweede Kamer. (2003). *Terugkeernota. Maatregelen voor een effectievere uitvoering van het terugkeerbeleid*. Vergaderjaar 2003–2004, 29344, nr. 1.

United Nations. (2000). Protocol against the smuggling of migrants by land, sea and air, supplementing the United Nations convention against transnational organized crime. http://www.uncjin.org/Documents/Conventions/dcatoc/final_documents_2/convention_smug_eng.pdf.

Dual Strategies

Chapter 13
The Civil RICO Law as the Decisive Weapon in Combating Labour Racketeering

James B. Jacobs

For much of the twentieth century the Italian-American Cosa Nostra organized crime families infiltrated and exploited American unions and those unions' pension and welfare funds. They exercised control by holding top union offices themselves and/or by exercising influence, even control, over certain union officials. Either way, they drained union treasuries through bloated salaries, no-show jobs, embezzlement, and fraudulent contracts with service providers. They sold out union members' contractual rights in exchange for employers' bribes. They established and policed employer cartels, often smoothing the way for organized crime cronies to take an interest in some of the cartel's member firms.

For most of the twentieth century, federal, state and local law enforcement agencies posed little opposition to this systemic criminality. While there were sporadic prosecutions, there was no national, regional or local political commitment to attacking organized crime. Under J. Edgar Hoover's nearly half century leadership (1924–1972), the FBI did not make organized crime control a priority. Moreover, until the late 1960s, evidence from electronic surveillance could not be introduced in federal courts. State and local law enforcement agencies lacked the expertise and resources to conduct long-term investigations and, in many cities, the organized crime families corrupted local politicians and law enforcement officials.

In 1968, Congress passed a comprehensive electronic surveillance law that provided for court-approved wiretapping and bugging[1]; henceforth, electronic surveillance would be the most important investigative technique for obtaining evidence against organized crime. In 1970, Congress passed two laws specifically aimed at combating organized crime: (1) the novel and powerful Racketeer Influenced and Corrupt Organizations Act (RICO) that provided powerful criminal, civil and remedial provisions; (2) the Witness Security Act that made it possible to protect witnesses, including Cosa Nostra members themselves, willing to testify against organized crime defendants. In 1972, Hoover died suddenly (of natural causes), opening the way for a new era in FBI history, an era in which organized crime control would soon be the agency's number one priority.

---

[1] By "wiretapping," I am referring to the interception of wire communication, i.e. telephone calls. By 'bugging," I am referring to eavesdropping in clubs, restaurants, homes and cars by means of listening devices planted in those buildings or vehicles.

D. Siegel and H. Nelen (eds.), *Organized Crime: Culture, Markets and Policies.*
© Springer 2008

As the 1970s progressed, the FBI assigned increasing numbers of agents, armed with electronic eavesdropping devices, to organized crime investigations. In 1975, Jimmy Hoffa disappeared in a manner and context that left little doubt that he had been the victim of a Cosa Nostra assassination. Hoffa was the best-known union figure in the U.S. At the time of his death, having served several years in prison on corruption and obstruction of justice charges, he had been recently pardoned by President Richard Nixon and was campaigning to regain his old position as General President of the International Brotherhood of Teamsters (IBT). The Hoffa assassination turned the attention of federal law enforcement to the role of organized crime in the labor movement. The ensuing campaign against labor racketeering became an important part of in the larger war against organized crime.

As is often the case in large organizations, innovation in organized crime control came from the periphery not the center. In the war on organized crime, the New York City FBI office led the way. From the early 1970s, that office had been investigating organized crime by means of squads focusing on each of the city's five organized crime families. After the Hoffa assassination, the five family squads made labor racketeering a key focus of their investigations. They came to understand that organized crime made much, if not most, of its money from labor racketeering and that labor racketeering was the bridge that smoothed the way for the organized crime bosses to move from the under world to the upper world.

The most important strategy in the anti-labor-racketeering campaign was the use of RICO's civil remedy to purge Cosa Nostra from organized-crime-dominated unions and to reform the governance and administration of those unions so that they would be resistant to future organized crime penetration. The campaign began with FBI and U.S. Department of Labor investigations in the late 1970s and with the filing of the first civil RICO complaint against corrupt union officials and their Cosa Nostra compatriots in 1982. The anti-labor-racketeering campaign continues to this day (summer 2006).

The RICO Law

In the late 1960s, at a number of hearings on the nation's crime problems, law enforcement witnesses told Congress that 24 Italian-American Cosa Nostra organized crime families were infiltrating legitimate industries, thereby corrupting the national economy. (That claim was somewhat inaccurate. The Cosa Nostra organized crime families were heavily involved in labor racketeering as early as the 1920s.)

The 1970 Racketeer Influenced and Corrupt Organizations Act (RICO) was meant to provide law enforcement officials with a powerful anti-organized crime weapon. RICO is a complex statute, defining four new crimes (three substantive offenses and a conspiracy offense) which carry very heavy prison terms, providing for mandatory forfeiture of the proceeds of criminal activity, authorizing civil suits by RICO victims for compensation, and authorizing the government to bring suit in equity to restrain future RICO violations.

First, RICO makes it a crime to use the proceeds of "a pattern of racketeering activity" to acquire an interest in an "enterprise,"[2] for example using money obtained through gambling, robbery or extortion to purchase an interest in a construction company. Second, RICO makes it a crime to acquire control of an enterprise by means of a pattern of racketeering activity. For example, it is a serious federal crime take control of a union or a construction company through extortion or fraud. Third, RICO makes it a crime to "conduct the affairs" of an enterprise through a pattern of criminal activity. Under this provision, it would be a RICO offense to use an enterprise, like a labor union, to extort labor peace payments from employers. Fourth, it is a separate and serious federal crime to conspire to commit any one of the three substantive RICO offenses.

A RICO defendant can be separately convicted and sentenced for up to 20 years imprisonment for violating each of the substantive offenses as well as the conspiracy provision. Moreover, the defendant can be additionally convicted and punished for each of the predicate crimes that constitute the pattern of racketeering activity. In addition, a defendant convicted of RICO must forfeit all the proceeds of his criminal activity.

RICO contains two civil provisions. One of them allows victims of a RICO offense to sue the perpetrator for triple damages. To my knowledge, this provision has never been used (for obvious reasons) to sue organized crime members. But the other civil provision has been used frequently and to great effect. It authorizes the U.S. Attorney General to go to federal court to obtain an injunction and other equitable relief to prevent a RICO offender from committing future RICO violations. The statute authorizes the federal courts to shape whatever remedy is necessary to prevent a defendant from committing future RICO violations. In pursuing this civil RICO remedy, the government need prove its case according to the civil (preponderance of the evidence) standard or proof. Under the civil rules of procedure, which apply to civil RICO litigation, the parties are entitled to wide-ranging pre-trial discovery. In contrast to criminal procedure, in a civil RICO case the government can take depositions from the defendants and other witnesses and obtain copies of all sorts of books and records. The court has the power to construct whatever remedy is necessary to prevent future RICO violations and may punish as contempt (carrying a jail term) noncompliance with its remedial orders[3].

Applying Civil RICO to Organized-Crime Dominated Unions

In 1982, federal prosecutors in New Jersey brought the first civil RICO suit against the officers of a union, International Brotherhood of Teamsters (IBT) Local 560. The government complaint named as civil defendants both union officers and

---

[2] All the terms are defined in the statute. "Racketeering activity" is a wide range of serious federal and state crimes. "A pattern" is two or more such crimes within a ten year period. "An enterprise" is any corporation, organization or association in fact, in short just about anything.

[3] Other examples of the application of the RICO case see in: Jacobs et al. (1994, 1998); New York State Organized Crime Task Force, Jacobs (1990).

organized crime leaders, members and associates. The lawsuit alleged that the head of the union was both a member of organized crime who had violated RICO by acquiring his union position through the commission of numerous crimes like extortion and bribery and that, once in control, he conducted the affairs of the union by means of a wide range of crimes including extortion and murder. The same charges were leveled at Local 560s general executive board and at organized crime members who allegedly aided and abetted the corrupt union officials. The government asked that the court remove the defendants from their union positions and appoint a trustee to reform the union's governance and to administer the union during the reform period. The civil trial, lasting almost a year, resulted in a big victory for the government. The judge ordered that the organized crime defendants have nothing more to do with the union and that the defendant union officers forfeit their union offices. In the future, union officers would be prohibited from having contact with the organized crime figures or with expelled members of the union. The judge appointed a trustee to supervise the rehabilitation of the union and ordered the union to pay for the cost of the trusteeship.

The remedial phase of the case proved to be much harder to resolve than the trial phase. It took 10 years for the trustee, aided by federal investigators and prosecutors, to wrest control of a union away from the Genovese Organized Crime Family. The entrenched clique of organized-crime-affiliated officials used all available means to stymie the trustee, including turning over control of the union to hand-picked successors who had no criminal records. The trustee returned to court many times to obtain supplemental orders removing some union officers and ruling others ineligible to run for office. The trustee hired an experienced union official to run IBT Local 560s day to day affairs and worked hard to win over the hearts and minds of the rank and file. Still, for every step forward there was a half step backward.

Surprisingly, perhaps, the rank and file union members did not universally greet the trustee and his staff as liberators. Many members did not know about the abuses in the union's management and were suspicious of the government's motives. The mainstream U.S. labor movement condemned the civil RICO suit, charging that the Reagan Administration was intent on destroying the free, democratic and independent labor movement. Federal investigators and prosecutors scoffed at these criticisms, arguing that is was cynical, even laughable, to equate call IBT Local 560 free, democratic or independent.

Ultimately, the court and the trustee prevailed. Many members of the clique allied with organized crime were expelled from the union. Finally, there emerged new leaders who condemned the old organized-crime-dominated regime. The trustee placed the pension and welfare funds on a sound footing with responsible trustees. After 10 years of remedial effort, the court-appointed union trustee and the federal prosecutors asked the court to dissolve the trusteeship.

U.S. v. IBT Local 560 showed how civil RICO could be used to attack the most virulent kind of entrenched criminality. Unlike criminal prosecution, even prosecutions of multiple defendants, the civil RICO strategy allowed the government and court to treat the diseased organization as a whole and over an extended period of time.

In addition, the cost of the remedial effort was shifted from the government to the union itself. In the years that followed, other federal prosecutors brought similar suits, all basically modeled after the Local 560 litigation.

U.S. v. IBT

In 1988, federal prosecutors in New York City (Rudy Giuliani's office) filed an omnibus civil RICO complaint against the general president and general executive board of the entire International Brotherhood of Teamsters and against many leaders of the Cosa Nostra organized crime families.[4] History may judge this to be the most ambitious effort to achieve organizational reform through litigation in American history. At the time, with more than 1.5 million members, the IBT was the largest private sector union in the United States. It was comprised of more than 600 local unions, like IBT Local 560. For many years the Teamsters had been notorious for being influenced by a coalition or Cosa Nostra organized crime families. In fact, previous litigation had proved that organized crime exercised substantial control over the multi-billion dollar Teamsters Central States Pension and Welfare Fund, the largest employer/union fund in the country. In effect, that fund for years functioned as a kind of bank for organized crime.

The lawsuit settled quickly 1 year after it was filed. The IBT defendants agreed to some sweeping changes, including a democratic election procedures for the union's top leaders. Of course, the union defendants promised to have no dealings with organized crime members or associates and the organized crime defendants agreed to play no role in the union's affairs. Most importantly, the settlement established a three person Independent Review Board (IRB) that would hear disciplinary cases against IBT officers and members and against IBT local unions as well. One of the Board's members was the former director of the FBI. The settlement provided that the IBT International would pay the full cost of the entire monitoring and remediation effort.

The IRB hired a former federal prosecutor to serve as head of an investigations office. Over the next 17 years, approximately 500 IBT officers and members were expelled (or "voluntarily" retired) from the union for associating with organized crime figures, failing to cooperate with the IRB or for committing various crimes against the union and its pension/welfare funds. The IBT conducted (with the trustee's supervision) three competitive elections for international officers. The incumbent regime is independent of the organized-crime-controlled clique that dominated the union from the late 1960s to the early 1990s. While substantial reform is clear, it is too early to say that every vestige of organized crime influence has been eradicated or that organized crime's influence could never reemerge.

---

[4] U.S. unions are called "international" when they have affiliated Canadian locals.

21 Civil RICO Suits Against Labor Racketeers

Altogether, the U.S. Department of Justice (DOJ) has brought 21 civil RICO suits against labor racketeers since 1982. Seventeen of the cases involve local unions and four involve international unions (IBT, Laborers International Union of North America, Hotel Employees and Restaurant Employees International Union and the International Longshoreman's Union.) The most recent lawsuit was filed in summer 2005 against the International Longshoreman's Association (ILA)and is pending at the time this article goes to press.

DOJ has never lost one of these cases. Only the IBT Local 560 case required a full trial to resolve. The rest of the lawsuits (excepted the pending ILA case) settled on terms favorable to the government because the defendants saw practically no chance of prevailing at trial given the mountains of evidence that the FBI agents and federal prosecutors had been able to accumulate from previous criminal and civil cases and current investigations. All of the settlements involved the appointment by the court of a trustee, monitor, or liaison officer empowered to enforce the decree and to report to the judge.[5]

The duties and powers of the trustees are defined in the court's decree or in the settlement documents. In some cases, the monitors have been authorized to administer the union. To carry out that responsibility, they have usually hired an assistant with experience in union administration. In other trusteeships, the trustee's role has been limited to investigating and bringing disciplinary actions against members and officers who violate the court's decree or the union's bylaws or constitution.

Experience has shown that both organizational reform and disciplinary actions need to be carried by the trustee. A previously organized-crime-dominated union cannot be reformed until the gangsters have been purged from the union. This requires investigative and prosecutorial expertise. In addition, however, the influence of organized crime cannot be completely expurgated until the union has been reformed. In most cases, by the time a civil RICO case is resolved, the union's treasury has been depleted and its administrative apparatus is in shambles. The trustee needs to move expeditiously to reform the electoral and administrative machinery, earn the support of the rank and file membership and encourage a new and independent group of leaders to step forward. There can be no effective reform as long as the rank and file members believe that the old corrupt clique will (or might) some day regain control of the union.

In some of the cases, the decree or settlement document limited the trusteeship to a specific term, sometimes as short as 18 months with the possibility of a short extension. Other cases, like the IBT Local 560 and the IBT International cases

---

[5] The title of the position designated to monitor/enforce the settlement or decree differs from case to case. Therefore, in some cases there is a "monitor," while in others there is a "trustee," "court liaison officer" or some other titled office. In some cases, it has been important to the defendants to avoid their monitor being called a trustee. Whatever the title, the powers and duties of the office are spelled out in the judge's decree or in the settlement document. It has often been necessary for these powers and duties to be subsequently clarified, even amended, by the judge.

discussed above, have led to trusteeships of indeterminate duration. Indeterminacy is preferable because a definite term of years will encourage the corrupt elements in the union to believe that if they lay low for a few years, they will eventually be able to return to power.

Most of the settlements have included provisions requiring the trustees to file with the judge periodic reports on the trusteeship's progress. The potential usefulness of these reports, in some cases, has been undermined by concern that no material intentionally or inadvertently compromise ongoing investigations. In addition, fearing that such reports will inevitably be made public, the trustees have been reluctant to discuss problems and suspicions. Thus, the value of the reports is not as great as one might have initially hoped. Nevertheless, they are a good starting point for anyone seriously interested in the challenge of reforming profoundly corrupted organizations.

Practically all of the trustees are private lawyers with prior experience as organized crime investigators and/or prosecutors. This has obvious advantages. Such individuals are very savvy about how organized crime families operate. They know what to look for and how to investigate criminal schemes. They know how to prepare cases for prosecution, or in this context, administrative disciplinary actions in front of the judge or court-appointed trustee (hearing officer). Perhaps most importantly, as former federal organized crime prosecutors, they are in a good position to enjoy the confidence of FBI and Labor Department investigators whose continued assistance during the remedial phase of the case is essential. Nevertheless, there are also disadvantages to appointing these former prosecutors as trustees. Because they are selected precisely because DOJ has confidence in them, they are susceptible to the criticism that they owe their jobs to cronyism and that their sizeable compensation is a kind of a reward from one friend to another. Such a view, loudly proclaimed by the civil RICO suit's critics, can reinforce or create cynicism among the rank and file union membership.

Moreover, what the former prosecutors don't bring to the table is expertise in union governance and administration. They don't know how unions are run, how grievances are resolved and how collective bargaining agreements are negotiated. Consequently, while they have been very successful in expelling organized-crime figures and their cronies from the union, they have been less successful, even unsuccessful, in changing the union's culture, i.e. in building a coalition of rank and file union members in support of reform and union democracy. Finally, because these lawyers are drawn from private law firms, they cannot devote all or even most of their time to the union's reform and the time they can devote comes at a very stiff cost. Poorly or modestly paid workers have frequently looked askance at the huge salaries paid to the trustees for their part-time work.

Is Civil RICO a Magic Bullet for Labor Racketeering?

Unfortunately, the government has made no effort to evaluate the more than two-decades-long experiment in using civil RICO to reform previously organized-crime-dominated unions. There exists nothing like a manual of best practices or even a repository of documents related to these cases. Incredibly, the men and women who have served as civil RICO trustees have never been brought together for a debriefing or even for a discussion of what works and what does not. My recent book, *Mobsters, unions and feds: The mafia and the American Labor Movement* (Jacobs, 2006) constitutes a first attempt to document and analyze these cases, but a comprehensive evaluation is far beyond the capacity of a single academic researcher.

In assessing the value of civil RICO as a strategy for attacking systemic organizational criminality, it is important to recognize that we are not dealing with an experiment that sought to compare criminal prosecutions with civil RICO in the labor racketeering context. To the contrary, there have been scores of criminal prosecutions during the same years as the civil RICO labor lawsuits. The criminal and civil suits often targeted the same individuals. Furthermore, it is also important to remember that the civil RICO union lawsuits were motivated initially by the desire to destroy the Cosa Nostra crime families, not to reform the unions. The criminal and civil remedial strategies have drawn on and reinforced one another.

With those caveats in mind, there is much to be learned from 25 years of civil RICO litigation against labor racketeers. First, civil RICO is perfectly suited to attacking systemic organizational criminality. It permits the government to target an entire organization for reform. Indeed it invites making the corrupted organization the target of remediation. Success should not be measured by the number of successful criminal and administrative actions against individuals, but by the extent to which the targeted organization runs freely, properly and efficiently (recognizing that those terms are not see easy to define much less measure).

Second, the government has had great success in bringing civil RICO suits against organized-crime-dominated unions. In large part this may be attributable to the fact that the government has chosen such egregious cases and marshal such overwhelming evidence (or at least allegations) that the defendants saw no chance of prevailing. But I think there is a deeper explanation. The civil RICO law itself gives extraordinary opportunities to government litigators. For one thing, the government does not have to prevail over every defendant, just over key defendants or enough defendants so that the judge (if it comes to a trial) will appoint a trustee to oversee the reform of the corrupted organization. For another thing, the structure of the RICO law allows the government to bring in extremely wide-ranging evidence about the nature and functioning of organized crime families as well as the operation of labor racketeering. So many allegations can be brought to bear in a single suit that there is virtually no chance of the defendants successfully fighting them all off. In all this, the government benefits from the civil procedures applicable to these suits.

Third, the civil RICO remedy invites and requires long term remediation. It is not a quick fix. The settlements that have been limited to a determinate time period have tended not to succeed. It has taken many years, perhaps in the range of 10 years, to liberate and reform the unions targeted by the civil RICO lawsuits. Those who bring the lawsuits must be ready for a long haul. They need to have a vision of what success will ultimately look like and then to develop a plan to get there. Organizational reform is not for the weak-willed.

Fourth, long-term remediation of a thoroughly corrupted organization requires a lot of resources. It is a major mistake to attempt such organizational reform "on the cheap." A competent trustee needs to be well remunerated. He or she needs staff and the ability to hire professional consultants – lawyers, auditors, accountants, loss prevention specialists, union democracy specialists and others. The judge or the parties themselves must assure that there is an adequate funding source for a multi-year effort.

Fifth, successful remediation of an organized-crime-dominated union requires the on-going participation of law enforcement and a working partnership between the remedial trustee and the law enforcement investigators. Much in the civil RICO complaint is backward looking. The civil complaint documents the history of corruption and racketeering in the union up till the present time; hopefully, from the government's perspective, that will be sufficient to persuade the judge to impose a trusteeship or to convince the defendants to save the expense of litigation and capitulate before trial. But the remedial phase requires information about organized crime's current influence in that union. Only the FBI, Labor Department investigators and other law enforcement personnel have or can get that information through informants, electronic surveillance and other investigative techniques. Their cooperation in the remedial phase of the lawsuit is essential. At a minimum, that requires that the trustee be a person whom the law enforcement agents trust completely. If they do not have that confidence, they will not share investigative information and the trusteeship will be doomed.

Sixth, much, probably too much, depends upon the creativity and skill of the trustee. Each of these civil RICO trusteeships has been *sui generis*. Some trustees have risen to the challenge and performed brilliantly; others have fallen short. This is obviously not a desirable state of affairs. We need a strategy that can be implemented by a large number of competent individuals, a strategy based upon sound conclusions drawn from prior trustee experiences. Thus, a strong evaluation component should be built into each trusteeship.

Conclusion

Civil RICO is the current generation's last and best chance to break the power that the Cosa Nostra organized crime families have exerted over many American labor unions for most of the twentieth century. There is good reason to be optimistic. Much has been achieved and much has been learned. However, much work remains to be done, especially the hard work of documenting and evaluating 25 years of civil RICO litigation against organized-crime-dominated labor unions and the identification and routinization of best remedial practices.

References

Jacobs, J. B. (2006). *Mobsters, unions and feds: The mafia and the American Labor Movement*. New York: NYU Press.
Jacobs, J. B., Friel, C., & Raddick, R. (1998). *Gotham unbound: How New York city was liberated from the grip of organized crime*. New York: NYU Press.

Jacobs, J. B., Panarella, C., & Worthington, J., III. (1994). *Busting the mob: U.S. v. Cosa Nostra*. New York: NYU Press.

New York State Organized Crime Task Force, & Jacobs, J. B. (1990). *Corruption and racketeering in the NYC construction industry*. New York: NYU Press.

Chapter 14
Recent Anti-Mafia Strategies: The Italian Experience

Antonio La Spina

Introduction: Indirect Strategies Against Mafia Type Organized Crime

We often tend to think of criminal policies as essentially repressive. At first sight this assumption seems to apply straightforwardly in the case of policies against organized crime, especially those targeted at mafia type organizations. However, experience shows that these criminal organizations can be fought successfully *also* by means of measures which are not in themselves repressive, albeit they aim to favour repression in the last resort. Rather than "alternative," I prefer to call such indirect strategies to fight organized crime "complementary" to repressive measures, in the sense that the latter are always required, but can be more effective if they are complemented by indirect means.

More than two decades ago, the Italian state introduced measures to fight mafia type organizations more effectively than was previously possible. The turning point was the Rognoni-La Torre act of 1982. From that moment onwards, the range of repressive measures expanded to a level which does not compare with any other European country. I am referring to specific criminal offences (the crime of mafia type criminal association); broad investigative powers; penetrating preventive measures (with regard to freedom of movement of suspected mafia men as well as to patrimonial assets and economic activity); regulations concerning financial transactions and money laundering; sanctions; special procedural rules on mafia trials, and so on.

Repressive measures are normally aimed at discouraging people from doing something by punishing them afterwards. Consequently, they will be effective if illegal actions are discovered, if the criminals are seized, if the trials are quick, if the benefits stemming from crime are reduced or eliminated altogether, if inmates cannot interact with one another and with the organization, and so on.

In addition to direct repressive measures, we can observe in the Italian case a wide variety of other policies addressing civil society and public administration, and only *indirectly* the mafia. I will illustrate such indirect ways to tackle the mafia problem in the next sections. Before doing so, I will first briefly elaborate on the meaning of indirect policies.

D. Siegel and H. Nelen (eds.), *Organized Crime: Culture, Markets and Policies*.
© Springer 2008

Direct policies are designed to punish people who act in a certain way, and to prevent them from performing or repeating such actions. In the case of mafia type organizations, it is possible that a given individual behaves in a certain way (e.g. swearing an oath of allegiance to the organization) which is not per se directly and immediately harmful to somebody, but can be punished anyway because it is the organization that is the target, not just the individual, or the individual's conduct. The mere existence of the organization in a given area exerts psychological pressure on citizens and entrepreneurs, who will adopt a certain attitude towards it. Some entrepreneurs may be induced to leave, some will resist the pressure, while others expect to be asked to give or do something for the mafia sooner or later. It is important to stress the difference between areas where the mafia is known to be present and considered a constant and pervasive property of the context (as is the case in some regions of Southern Italy), and areas where mafia type organizations may be active, but only on a selective and occasional basis. Another possibility is that the pressure of mafia type organized crime is limited to certain ethnic groups or certain activities.

When the mafia's presence is pervasive and structural, this either inhibits repressive action (which was the Italian situation before the turning point), or eventually makes for the growth of direct repressive policies aimed at eradicating these organizations. At the same time, a wide range of indirect policies will also be required. Conversely, if the mafia's activities are occasional and circumscribed, there is less need for indirect policies.

Consider a context like Sicily, Calabria or Campania. If most entrepreneurs and shopkeepers refused to pay extortionists but rather reported them to the police, if citizens stopped voting for political figures who are known to be directly or indirectly linked to the mafia, if eyewitnesses decided to give their testimony regardless of retaliation, if all the "white collars" (professionals, civil servants, politicians, entrepreneurs, opinion makers) avoided favouring the mafia in any way, if enough ordinary citizens engaged in grassroots anti-mafia activities, and if enough mafia men chose to collaborate with the police and public prosecutors, then after some time the mafia would disappear, because its members would frequently be exposed and apprehended, and the organization would lose its sources of income and eventually collapse.

To expect such a change in social conduct to happen overnight, is not realistic. In recent times there have been many encouraging signs but most people still do not rebel against the mafia because the price for doing so is too high. If the balance between costs and benefits is not altered, only a few exceptional people will resist the mafia consistently through time. On the other hand, it still happens in Italy that people (entrepreneurs, politicians, etc.) cooperate with the mafia not on the basis of fear, but rather because such actions produce an illegitimate gain. Cooperative strategies like this should be discouraged and punished.

Indirect strategies are aimed at reinforcing and spreading attitudes and behaviours which can pose serious obstacles to the day-to-day activities of mafia men. Most of the time, these strategies do not concentrate on professional criminals. It is one thing to capture individual fish, but it is quite another to drain the water where they swim, or to create systematic hindrances to their movement. Although repression is still the main feature of Italian anti-mafia measures, indirect instruments are becoming more and more important. In what follows we will illustrate the main measures based on an indirect approach and their rationale, as well as some other measures clearly aimed at repression but with indirect components.

Indirect policies are rather heterogeneous. They address a wide range of actors: entrepreneurs, civic associations, public administrations, local bodies, schools, and young people. For instance, they try to promote a "culture of legality" and a reaction by civil society against the mafia; they render public administrations and public acts more impermeable to mafia influence; or strengthen the capabilities and performance of police forces. Sometimes they aim at spreading information, awareness, and motivation against the mafia; sometimes they try to alter the balance of costs and benefits from the point of view of certain actors. This is the case when it comes to anti-racketeering legislation and legislation for eyewitnesses and collaborators of justice. Most of the time, they try to *promote*, to *favour* or to *induce* a behaviour which is normally rare and costly (albeit in some cases such actions are performed spontaneously).

Anti-Racketeering Actions

Anti-racketeering legislation aims at restoring various kinds of costs borne by people who resist the racket. The first act in this field was adopted in 1992, after the murder of the entrepreneur Libero Grassi, who had publicly announced his decision to refuse paying the racket. The act was designed to reimburse some of the damages suffered by people who resist extortion. In the same period some small firms began to create anti-racketeering associations. A well known leader of this movement, Tano Grasso, a shopkeeper in the small town of Capo d'Orlando, wrote that "the association is the instrument that can guarantee safety to those entrepreneurs who decide to denounce the racket. There is a need to be many ... we must assert the idea that there is a commonality of interests which links the direct victim of a given extortion to all other economic actors. The fact that individual actors resist is not sufficient if they are alone" (Grasso, 1992: 78).

A new legislative act adopted in 1999 reformed the previous measures, which had proved largely ineffective, given that almost all the allocated financial resources remained unspent. The 1999 act increased the amount of benefits granted to racket victims and their relatives by extending the types of reimbursable damages, made their payment quicker and more reliable, involved the associations in its management, eased the burdens of cooperation between victims and investigators, and created an ad hoc Commissioner, who is also responsible for coordinating actions against usury. The act recognizes that the choice to oppose the racketeers and to denounce them is not an easy one. Therefore, it considers the possibility that somebody who at first refused to pay might subsequently change his views. In this case, the benefits that he or she might have received would be revoked.

The first anti-racketeering Commissioner was the already mentioned leader of the anti-racketeering movement Tano Grasso. He acted to speed up the processing of requests and the subsequent payments, as well as the diffusion of information about the measure.

However, from 1996 onwards, the number of denunciations for extortion did not grow, but rather decreased, and this trend was not influenced very much by the new act. In some cases, denouncers and their families were forced to change their lives completely in order to receive constant protection. Banks, possible clients and suppliers sometimes see people who resist the racket as too risky to deal with. Generally speaking, the compensations offered by the 1999 act still appear to be insufficient to bring about a general change in attitude.

A repressive approach (instead of a promotional one) might also be used in order to induce (or, better, to force) entrepreneurs to cooperate with the police and the judiciary. According to some proposals which have been the subject of debate in Italy, it should be considered a crime in itself to pay extortionists (an act which, in most of the cases, is to be regarded as a sign of victimization), because it helps the organization, confirms its power and does not support repressive action. Presently, this is not the case, but when instances of extortion are actually discovered and a victim is afraid to cooperate with the enquirers, in some cases he or she can be charged with abetting.

New policies are presently debated which are aimed at restoring damages suffered as well as realizing additional benefits for firms that oppose the racket. One of the possibilities under consideration is that of preferential treatment of these firms when it comes to public procurements or when the concession of public services or State property is at stake (Ministry of the Interior, 2003).

In more recent times, there have been some other interesting novelties. In 2004, local organizations of entrepreneurs proposed to put a "blue stamp" on the goods produced by firms who are ready to submit themselves to controls against money laundering. In the same year, a new movement of students was created in Palermo, called "Addiopizzo." "*Pizzo*" is a slang term referring to the sum of money that the racket periodically asks firms to pay, and the name of the association means "goodbye extortion." Its first actions were mainly of a symbolic nature: anonymous stickers or banners with phrases like "a people who rebel against the mafia and refuse to pay the *pizzo* is a free people." In May 2006, a new initiative was taken: a list was published of over one hundred shopkeepers and entrepreneurs in Palermo who all stated that they were prepared to resist the racket. The number is perhaps relatively low compared to the overall number of entrepreneurs in the city, but it is hopefully enough to produce the protective effect to which Grasso referred in the passage quoted above. This initiative creates a "safety belt," sets an example to be followed by other people, and, last but not least, invites consumers to "reward" these economic actors by opting for their products and services. In this way, with the help of a grassroots movement, an additional economic benefit is generated which could make the difficult choice to resist the mafia at least somewhat easier.

Public Administration

The sector of public works and procurements is an area where the mafia traditionally tries to determine the winners, by organizing "queues" of firms who cooperate with each other with regard to tenders, in order to coordinate the bids and influence the result, so that all of them get their turn at winning a contract. This normally requires a favourable attitude on the side of political and administrative decision-makers as well. Before the period known as *Tangentopoli*, Italian legislation on public works and procurements followed a different approach. Corruption was widely practiced, albeit not always in connection with the mafia. In 1999, after a halting start, the Authority for the supervision of public works (*Autorità per la vigilanza sui lavori pubblici*) began its activities, with the objective of guaranteeing legality, efficiency and accountability in the sector. This independent commission can inspect and sanction firms, as well as ask for documents and information required by the public bodies concerned. It must then denounce to the judiciary any irregularity detected. In 2003, a co-ordination Committee for the high supervision of major public works started its work. Its members are representatives of the Ministry of the Interior, the Ministry of Infrastructure and Transportation, the DNA (*Direzione nazionale antimafia*, a public prosecutor with national competence), the Authority for the supervision of public works, the DIA (*Direzione investigative antimafia*, a national specialized police body), and the Central Directorate of the Criminal Police. Its task is to prevent mafia type organizations from penetrating major infra-structural projects.

Some regions with special autonomy (notably Sicily) have their own legislation concerning public procurement. At the beginning of the nineties it was reformed in such a way as to avoid discretion, to concentrate procurement operations, and to make them impermeable to external pressures. It was then reformed again, always making resort to "automatic" procedures (not to maximum reduction, but rather to a complex mechanism based on the average of the bids).

It is difficult to evaluate the effectiveness of the legislation on public works and procurements in avoiding mafia type infiltration. The above-mentioned cooperation between the Authority for the supervision of public works, the DNA, the DIA, and the relevant ministries looks promising. The different solutions experienced in Sicily were disastrous. The first version of the relevant regional statute produced blockages in the public works department. But its second version introduced an "automatic" mechanism which enabled mafia men to intervene, by managing "chains" of entrepreneurs able to determine the results of bidding procedures. This statute was recently reformed again, imitating national legislation.

Firms or individuals dealing with the public sector (applicants for public assistance or companies bidding for public contracts) were requested (for the first time in 1982) to submit an "anti-mafia certificate" attesting that they had no involvement in mafia type associations, which meant not only that they had never been convicted of mafia related lawbreaking, but also that they were not subject to the preventive measures. This administrative burden was subsequently simplified and mitigated. Besides the private subject, it is now also the public administration body concerned which asks the prefecture about the situation of the proponent.

Given that mafia type organizations are frequently able to influence local politicians, there are rules which allow for the suspension or dismissal of mayors or presidents of regions and provinces, as well as for the suspension or dissolution of provincial and municipal councils in cases where a connection between local administrators and the mafia can be shown. These rules address political personnel only, while an appropriate treatment of the compulsory transfer or expulsion of civil servants colluding with the mafia is still lacking. We must also keep in mind that such measures are enforced by the national executive in charge, which will have a certain political orientation. Actions against local mayors and councils of the same orientation can put the relevant parties in an awkward position, while if the local authorities are politically affiliated to the opposition, the national majority could be accused of being biased. As a matter of fact, such measures were applied more frequently in the first half of the nineties, and less frequently afterwards. From 2001 onwards, there has been a revival of this type of disciplinary actions.

Both local councils and the subnational level of government (called a "region" in Italy) are increasingly committed to anti-mafia action. An indicator of this is the fact that in the Southern regions, where the mafia is endemic, homicides of local politicians and, more often, intimidation against them has happened much more frequently in the last 5 years than in the past. Some local councils, provinces or regions have enacted ad hoc provisions to help anti-racketeering and anti-usury movements (one example is that of the "anti-usury consulting rooms" to give legal advise and psychological support to victims of usury (Grasso, 1998), and have joined as plaintiffs in criminal lawsuits against mafia men.

Indirect Features in Measures Concerning Collaborators

An important policy development, especially from 1990 onwards, has to do with promoting, rewarding and using "collaborators of justice," who are either private witnesses, victims or, more significantly, people formerly involved with the mafia who have decided (usually once they are in jail) to abandon the organization and to cooperate with public prosecutors and the police (the so-called *pentiti*). Such a policy, which proved very successful, is closely related to repressive action, because collaborators can be of help in the apprehension, prosecution and conviction of mafia men, as well as in shedding light on their covert businesses and properties. Furthermore, collaborators are frequently mafia members who are already in jail, possibly under a strict incarceration regime. Therefore, the choice to collaborate can be a way to avoid the costs of repression. On the other hand, many mafia members who are in jail choose *not* to collaborate, because they prefer to continue to belong to the organization, sticking to its "code of honour," and/or because they fear retaliation from the organization against themselves or their relatives. In other words, there is one similarity between the otherwise totally different situations of the entrepreneur who must be persuaded by anti-racketeering regulations to denounce his extortionists and that of the mafia man who must be persuaded to "repent": the fact that society wants them to do something while they have strong incentives *not* to do so. In my view, the promotional nature of the measures regarding *"pentiti,"* together with the fact that they imply several administrative aspects, puts them halfway between direct and indirect strategies.

Apart from its contribution to repression, the very act of repentance is in itself a menace to the foundations of the organization: secrecy, mutual protection, and total identification between the life of the individual and his being a member of the organization (Paoli, 2003, mentions the "contract of status" which defines the identity of the man of honour). Of course, this applies to mafia type organizations and terrorist organizations, but not to any other kind of organized crime. It was not by chance that the sanctions adopted by the mafia against the first *pentiti* and their relatives were devastating. To this day, collaborators and their families need strict and continuous protection. For this reason, a protection program is proposed for each new collaborator to an ad hoc Central Commission. The program is then managed by a central Protection Service. Such programs also foresee that the collaborator and his relatives change their residence; that he is held in custody according to a special and favourable regime (including the possibility to benefit from the various "alternatives" to jail); is covered through new identity documents; and obtains substantial penalty reductions, as well as financial assistance. When the collaborator is heard by the prosecutor and during the trial, specific precautions are taken, such as the utilization of videoconferencing to allow collaborators to testify without leaving their safe residences.

The rules on collaboration have raised a considerable amount of criticism. First of all, many people found it difficult to accept that criminals guilty of several homicides and other serious crimes were sometimes released because of their repentance. Secondly, it was at least possible that collaborators were manipulated, as well as it was possible for seasoned criminals to "manipulate" public prosecutors by means of partial, incorrect or false declarations, especially if cross-checks with hard facts were lacking. Indeed, some decisions of the Court of Cassation recognized as valid and sufficient evidence the "cross-confirmation" of the depositions of different collaborators, even in the absence of "objective" counterparts. Other decisions deemed admissible a *de relato* declaration, which is information based not on what the collaborator says he has directly experienced, but rather on statements supposedly made to him in the past by people now dead, missing, or unwilling to confirm.

Act 45/2001 amended the existing rules, with the following aims and issues in mind: the promotion of the cooperation of witnesses not associated with the mafia; the need for the collaborator to declare illicitly obtained assets so as to make his testimony more credible; the necessity of cross-examination during the trial; the requirement that the collaborator serves a minimum portion of the sentence imposed by the court; the avoidance of coordinated statements between him and fellow collaborators or other parties concerned; the restriction of "declarations by instalments," given that several years after their initial "repentance," some collaborators suddenly "remembered" that it would have been appropriate to accuse people they never even mentioned in their previous statements. These days, a collaborator has to state what he knows within the 6 months following his decision to cooperate. This period is considered too short by many public prosecutors. Still, in his evaluation of the new statute, the general prosecutor of the Supreme Court of Cassation (who speaks on behalf of the whole judiciary at the opening of the judicial year) said that it "should make the phenomenon of collaboration more effective from the trial's point of view, as well as more socially acceptable, so dispelling the thick halo of mistrust that surrounds it" (Favara, 2002).

In 1997 the number of collaborators decreased sharply, because the Central Commission adopted strict criteria for admitting new collaborators. This means that there were probably too many before that period. When statute 45/2001 was passed, many judges feared that it would lead to a further decrease in new collaborations. But the number of pentiti remained more or less the same as it was in 1998, 1999, and 2000. In recent years, a number of new collaborators have contacted the authorities. A strong incentive to collaborate can be found in the especially severe incarceration regime applicable to mafia members, a fate pentiti hope to avoid. It must also be emphasized that recently several important figures have chosen to collaborate, one Sicilian example being that of Giuffrè in 2002. In addition, several mafia women and, more recently, a young politician named Campanella have also decided to cooperate with the police.

The management of collaborators can be very costly, because they need extensive protection and a convincing incentive to cooperate. On the other hand, the advantages they have generated for the anti-mafia fight have been, and still are, considerable. Not only have they provided prosecutors with crucial information, but they have also created an atmosphere of general insecurity and mistrust, forcing mafia men to change their standard operating procedures, which are now much more costly, risky, and less visible. What is significant is that through the interception of communications it is now much easier than before to find objective evidence to support the testimony given by collaborators. In the fight against mafia type organizations, the appropriate use of collaborators – in itself a slippery and risky instrument – can be more than worthwhile, and is therefore justified.

Strengthening the Effectiveness of Police Operations

Another strategic move which is not in itself repressive, but which makes prevention, detection, and repression of organized crime much more effective has to do with strengthening the technological capabilities of the police forces. Almost all of the Mezzogiorno (and notably those regions where mafia type organizations are pervasive) falls within the so-called Objective 1 of the European cohesion policy. For the period 1994–1999, a National Operational Program (PON) was launched, called "Security and Development for the Mezzogiorno" (SDM), with a grant of around € 280 million. The aim was to improve security standards and the rapid response of flying squads through an integrated system of satellite telecommunication, technologically up-to-date sensors located in the area, and operative interconnected control rooms. Other projects concerned the strengthening of national border control, as well as technological and human resources development in fields such as infrastructural networks for radio communications, identity verification, x-ray control equipment, and so on.

For the period 2000–2006, the PON SDM was granted funding of around €1120 million (Public Security Department, 2000, 2003). In this second phase, it covers all the territories of all Objective 1 Italian regions. The overall aim is to produce and stabilize security standards in the Mezzogiorno equal or at least similar to those that can be found in the rest of the country. More specific objectives are, among others, a further strengthening of the use of new technologies, the inducement of quicker responses and decisions by police forces and the courts, the diffusion of a culture of legality, a closer involvement of civil society, the use of technology for the protection of environmental and cultural resources, and social and economic development at the local level. The PON SDM is the only program in Europe which couples the aims of socio-economic development with that of the security of citizens. It is already regarded as a "best practice" approach, and its "transferability" to other countries, particularly the Eastern European members of the enlarged European Union, was taken into consideration by the DG Justice and Internal Affairs of the European Commission (PON SDM, 2003).

A targeted and effective use of new technologies can be devastating to mafia activities. "Men of honour" usually speak very little, but they cannot avoid a certain amount of crucial communication, both between themselves as well as with their victims. In the past (and often still today) the mafia moved without anybody seeing or hearing anything, but if phone calls and conversations can be tapped and acts can be filmed, there is no need for witnesses anymore, and the work of mafia men becomes much more difficult. As is well-known, Bernardo Provenzano – the boss of all bosses in Cosa Nostra – used an archaic medium (small slips of paper, so-called *pizzini*), to avoid interception. But even small sheets need "postmen" to deliver them and postmen can be detected. Provenzano was caught in April 2006, after having spent more than 40 years on the run, because the police were able to monitor his relatives and the people who brought him clean underwear.

There have been other interesting cases. Several entrepreneurs decided to denounce their extortionists and allowed themselves to be filmed with the aid of micro-cameras while they were paying the *pizzo*. In other, more recent cases, prosecutors managed to obtain objective evidence of extortion without the aid of the victims. When the mafia wanted to participate in the management of a well-known restaurant in the historical centre of Palermo, the *Focacceria San Francesco*, and pressed the owners with requests for money and proposals that were "hard to refuse," investigators suspected what was going on and put the place under surveillance using bugging devices and undercover policemen. Eventually, they were able to catch several mafia men, while the owners of the restaurant continued for some time to deny that they were being blackmailed. Developments such as these might partly obviate the need for collaboration by victims of the mafia and mafia men alike.

New technologies and the necessary training can be costly, not only in financial terms, but also because they can put at risk the privacy and the freedom of persons not connected to the mafia. Much like public opinion in other countries on the subject of terrorism, the Italian public apparently believes that some infringements on personal liberties are justified, provided that they serve the fight against the mafia, and will not be applied to other realms of society.

The Spreading of a "Culture of Legality"

A number of activities have been set up to support collective anti-mafia movements and to sensitise younger generations both to the threat posed by organized crime as well as to their own civic duties. Schools and universities, with the help of local councils and regions, are frequently engaged in projects concerning the so-called "culture of legality" (Santino, 2000, 301 ff.; Schneider & Schneider, 2003, 260 ff.).

Italian legislation provides for the rapid seizure of valuable real estate and the targeting of "straw man" ownership (even outside of the family circle) on the basis of collaborators' declarations or intercepted communications. These assets are extremely significant, although there are serious difficulties in appraising the properties and assigning them to alternative uses. Confiscated real estate can be used by local councils and by non-profit organizations for social ends. The symbolic impact of setting up schools, social centres, or other public facilities using resources formerly belonging to men of honour is understandable. The mafia once deprived the community of this wealth, but now these very properties can render a service to the same community. Nevertheless, the procedure leading to public utilization is slow and uncertain. As a consequence, the economic value of some of these objects diminishes or vanishes altogether. Properties can fall into disrepair, and the municipalities or non-profit organizations which are supposed to put them into use may find this too costly because of the necessary restoration expenses.

In some cases, confiscated mafia property is used for legitimate economic activities. "Legality and Development" is a consortium between several municipalities in the Palermo province, including Corleone (the hometown of Liggio, Riina and Provenzano), which aims at generating occupational opportunities through the use of confiscated goods as well as spreading the culture of legality. The productive sectors involved are agri-tourism, organic farming, and zootechnology. One example is a co-operative society called "Placido Rizzotto – Libera Terra" (named after a trade-unionist killed by the mafia), which labels its produce as biological products coming from land that was "liberated from the mafia" (to be precise: some of this land once belonged to Totò Riina).

In 1995, a national alliance of associations was created, called "*Libera. Associazioni, nomi e numeri contro le mafie*" (Free Associations, names and numbers against the mafias). It is led by a Catholic priest named Luigi Ciotti. Libera designs and manages projects related to the use of confiscated assets; it selects and trains personnel; and organizes volunteers.

Another instrument used at the local level has to do with the experience of "bargained programming" to promote socio-economic development at the local level, involving municipalities, other public bodies, entrepreneurs and their organizations, and social partners. Some of these agreements include "legality protocols," with the participation of the prefect and the police forces, through which the parties assume reciprocal obligations concerning the strengthening of security conditions and public order in the area (with regard both to economic activities and to the community level). Financial support can be obtained from a national Security Fund.

There are also "agreements on security," which consider more specific measures against the various types of crime (including micro-criminality) through widespread control of the territory, so as to foster a sense of safety in the citizens and to make the areas concerned more attractive to external investors.

Finally, anti-mafia initiatives institutionalized "from above" must be mentioned. I am referring to the many centres, associations, and foundations named after victims of the mafia, which often receive financial support from regions and local administrations. Most of these develop programs (together with schools, universities and actors within civil society) with the aim of spreading opposition to mafia type crime and enhancing academic understanding of these issues. In some cases there is a risk of repetition and ritualism.

Concluding Remarks

A few years ago, a book came out entitled "Why my name is Giovanni" (Garlando, 2004). The author is a father who tells his son why he bears the same name as Giovanni Falcone, one of the heroes in the fight against Cosa Nostra. In simple language, the father explains what the mafia is, how it evolved, and why we need to get rid of it. A book of this sort is meant to express the feelings of the Sicilian population and it is only one of many examples of an anti-mafia consciousness. All over Italy, especially in Sicily, Calabria and Campania, schools are engaged in educating young people in understanding the roots and dimensions of the phenomenon and teaching them how to resist it. Nevertheless, it is difficult to tell the exact number of people who resist the mafia on all fronts, and academic evidence obtained through appropriate surveys is presently lacking. Today in Sicily, Campania, Calabria and Apulia many entrepreneurs still resign themselves to being victims of extortion, and significant actors in the fields of politics, economics, public administration, and the professions still find it convenient to collude with the mafia. Therefore, Italian indirect anti-mafia policies are still a long way from final victory. Direct policies, on the other hand, are now much more effectively implemented than they were in the past, especially with regard to the Sicilian mafia (Cosa Nostra) as is shown, for instance, by the number and stature of those members of the organization now in jail; by the quantity of assets confiscated; and also by the fact that some of the leaders of the organization have offered the Italian state a "peace treaty" (for a review of the available evidence see La Spina, 2004, 2005). When the public recognises the effectiveness of these policies, this can be a great help in fostering opposition by civil society, economic actors and local communities against mafia type criminal organizations. When the people realise that the mafia men are under stress, they will be more ready and able to resist.

All this means that direct as well as indirect anti-mafia policies in Italy must be sustained and reinforced until the mafia is finally defeated. It is possible to envisage a European policy aimed at promoting the reaction of civil society against the various forms of mafia type organized crime, which is of course by no means confined to

Italy. However, in European countries where the phenomenon is not as widespread and challenging as it is in Southern Italy, risky methods such as the massive interception of communications and the use of collaborators of justice (which should be used with the utmost caution in Italy as well) can hardly be legitimized.

References

Favara, F. (2002). Relazione sull'amministrazione della giustizia nell'anno 2001 del procuratore generale della repubblica presso la Suprema Corte di Cassazione, 11.January 2002 (http://www.giustizia.it/).

Garlando, L. (2004). *Per questo mi chiamo Giovanni*. Milano: Fabbri.

Grasso, T. (1992). *Contro il racket. Come opporsi al ricatto Mafioso*. Roma-Bari: Laterza.

Grasso, T. (1998). L'ambulatorio antiusura. Economia, legalità, criminalità. *Strumenti, studi e ricerche, I*(2), 11–30.

La Spina, A. (2004). The paradox of effectiveness: Growth, institutionalization and evaluation of anti-mafia policies in Italy. In C. Fijnaut & L. Paoli (Eds.) *Organised crime in Europe: Conceptions, patterns, and policies in the European Union and beyond* (pp. 641–675). Dordrecht: Springer.

La Spina, A. (2005). *Mafia, legalità debole e sviluppo del Mezzogiorno*. Bologna: Mulino.

Ministry of the Interior. (2003). Rapporto annuale sul fenomeno della criminalità organizzata per l'anno 2002. Parte III. Strategia e azione di contrasto. Roma. http://www.interno.it/.

Paoli, L. (2003). *Mafia brotherhoods: Organized crime, Italian style*. Oxford: University Press.

PON SDM. (2003). Obiettivo Sud, Newsletter, 1–2, January–February, http://www.sicurezzasud.it.

Public Security Department, Ministry of Interiors. (2000). Programma Operativo Nazionale Sicurezza per lo Sviluppo del Mezzogiorno, Quadro Comunitario di sostegno, Italia, Regioni Obiettivo 1, 2000–2006, July 2000, http://www.interno.it/dip_ps/.

Public Security Department, Ministry of Interiors. (2003). Programma Operativo Sicurezza per lo Sviluppo del Mezzogiorno d'Italia, Complemento di programmazione, programming period 2000–2006, as approved by the Surveillance Committee in January 2003, http://www.interno.it/dip_ps/.

Santino, U. (2000). *Storia del movimento antimafia*. Roma: Editori Riuniti.

Schneider, J., & Schneider, P. (2003). *Reversible destiny. Mafia, antimafia, and the struggle for Palermo*. Berkeley-Los Angeles: University of California Press.

Chapter 15
Breaking the Power of Organized Crime?
The Administrative Approach in Amsterdam

Hans Nelen and Wim Huisman

Introduction

In 1996, the Dutch criminologists Fijnaut and Bovenkerk reported to a parliamentary committee on police investigation methods in organized crime cases that Amsterdam had to be regarded as a "centre" for both national and international organized crime (Fijnaut & Bovenkerk, 1996). Some of the criminal groups involved had allegedly built up economic positions of power in the hotel and catering sector, the gambling sector and the property sector, mainly in the inner city districts and especially in the red-light district.

With regard to the red-light district, the researchers concluded that "criminal individuals and groups have, through their illegally acquired property and capital, gained control of most of the economic power. As a result, this enables them de jure and de facto to decide who, and to what extent, can develop illegal and/or legal activities, and thus, to a high degree, ultimately determine the level of public disorder or order in this area" (Fijnaut & Bovenkerk, 1996: 126). The report stated that the indecisiveness of the local authorities had created a fertile breeding ground for illegal and criminal activities in the red-light district. In particular, Fijnaut and Bovenkerk anonymously referred to sixteen "criminal" entrepreneurs who had allegedly become the key players in the red-light district.

In reaction to these findings, the city of Amsterdam decided, amongst other things, to place a greater emphasis on an administrative approach, in order to prevent the facilitation of organized crime. A central premise of this approach is that public administration services or facilities are needed to carry out criminal activities or invest illegally acquired capital. When criminal organisations can be excluded from public contracts or from receiving subsidies or licences for certain activities, the investment of criminal capital and the infiltration of the legal economic sectors will, to a large extent, be hindered. For this purpose, administrative bodies have become involved in combating a form of crime that had previously been the sole reserve of the police and the judiciary.

The concept of an administrative approach to organized crime was developed in the 1980s and 1990s in New York City by the former Guiliani administration. In addition to and in close coordination with a very intensive criminal policy – partly

D. Siegel and H. Nelen (eds.), *Organized Crime: Culture, Markets and Policies.*
© Springer 2008

based on the Racketeering Influenced and Corrupt Organizations Act (RICO) –
administrative measures enabled the authorities to break the positions of
power of the five big Cosa Nostra families in the city (Jacobs, Friel, & Radick,
1999).

Naturally, the Amsterdam administrative approach has a different background and has to be judged within the specific local context of the Dutch
capital. This approach consists of a number of instruments, ranging from the
integrity testing of civil servants, the purchasing of strategically-positioned
buildings and the refusal or withdrawal of permits, to the screening of companies who compete for major contracts. Because of the specific problems in
the red-light district, members of Amsterdam city council emphasized that
additional administrative attention should be devoted to this part of the city
centre and called for the appointment of a so-called red-light district manager.
On 1 February 1997, the city of Amsterdam did indeed appoint a manager
for the red-light district. In 2000, the name of the project was changed to
the *Van Traa project* and its scope was extended to the city of Amsterdam as
a whole.

The administrative approach to organized crime is in line with recent
developments in the context of situational crime prevention. Intervention strategies
aim at eliminating opportunities for committing crime and money laundering.
The innovative characteristics of the administrative approach in Amsterdam
and the Van Traa project in particular, raise two important questions: how has
this project been applied in daily practice and can it be regarded as an additional instrument which is effective in preventing organized crime? To answer
these questions an evaluation study was conducted. The final report of this
study was published last year (Huisman et al., 2005). In an article in *Crime, Law
and Social Change*, both the policy theory and the daily practice of the Van Traa
project were recently clarified and reviewed (Huisman & Nelen, 2007). In this
contribution, emphasis will be placed on the outcome of the project. Are there any
plausible effects – and side effects – of the Van Traa project on organized crime in
Amsterdam? Before addressing this issue, the development of the administrative
approach in Amsterdam will be described.

The Administrative Approach to Organized
Crime in Amsterdam

Three interrelated projects form the backbone of the administrative approach to
organized crime in Amsterdam. The first project is linked to the Integrity
Bureau, which was established in 2001. This bureau is mainly responsible for
activities of an internal organisational nature. The main objective of the
Integrity Bureau is to develop and promote a municipal integrity policy and to
monitor the developments in this area. The second element of the administrative approach is formed by the activities of the Bureau for Screening and

Auditing (abbreviated to *SBA* in Dutch), which operates under the direct authority of the mayor. This bureau is responsible for screening and monitoring all parties involved in tender procedures of large infrastructural projects in relation to construction activities, communication, data transfer and so forth. The third integral element of the administrative approach is the Van Traa project. As mentioned above, in 1997 a red-light district manager was appointed at the request of the city council, with the objective of improving the prevention of organized crime in the red-light district. The red-light district manager and his team were asked to develop a methodology for the administrative approach to organized crime. The Van Traa project was launched in 2000. Since then, the methodology developed by the red-light district manager has also been applied in other city districts and in specific economic branches. Examples of the selected areas are run-down streets in poor areas with a high number of immigrants, the most expensive shopping street in the city and the industrial harbour district. Examples of the selected branches are the so-called *smart shops* which could be participating in drug trafficking, phone centres which could be used for informal value transfer systems and the escort business which could be used for trafficking women.

The Van Traa project is intended to have a multi-agency approach in which several agencies cooperate by sharing information and integral enforcement. A small team – the Van Traa team – coordinates all activities. Nowadays, besides a team manager, the team consists of several legal advisers, project leaders and information specialists. Besides using its own methodology, the Van Traa team also coordinates the implementation of the Public Administration Probity Screening Act, hereafter referred to as the "BIBOB" Act, on behalf of the city of Amsterdam. The BIBOB Act, which came into effect mid-2003, enables the refusal or withdrawal of certain favourable decisions, and the refusal of bodies to participate in public tenders or contracts. This is applicable if there is a serious risk that a favourable decision will also be used to commit criminal acts or to utilise any financial benefits which have been or may be gained through criminal activities.

The methodology that has been developed in the framework of the Van Traa project comprises of two components: firstly, the collection and analysis of information about the selected areas or branches, and secondly, the taking of measures on the basis of this information. The team is given access to all the relevant information from the local authorities on the housing situation and the use of real estate in the selected areas and branches. In every subproject, several steps are taken to gain a clear picture of the crime in the district or sector concerned by consulting and linking information from public sources, municipal records and, if appropriate, classified information from the police, judiciary and tax authorities. With regard to the latter, the project team is given special authority by the Minister of Justice to access classified police information. By means of combining and analyzing all these sorts of information, an assessment can be made of the involvement of criminals in the selected area or branch. The second step is to take measures on the basis of this assessment. Because different partners work together in the Van Traa project, a wide range of measures can be taken, varying from the refusal or withdrawal of licences and permits, the levying of taxes, the closure of certain establishments, the initiation of criminal investigations, and, under certain circumstances, the acquisition of real estate by the city itself, in order to prevent criminals from investing their money in specific objects. In this respect, it is better to refer to this approach as a form of integrated maintenance, rather than the administrative maintenance of public order.

Multi-Agency Approach

Cooperation is a very important condition for an administrative approach to be effective. As much as possible, information has to be shared and exchanged, and measures which are taken as part of the intended integration of the maintenance of law and order have to be harmonized.

Besides the city districts and the city departments, the most important partners are the police, the Public Prosecutor's Office and the tax authorities. The private partners involved are several housing corporations and two private real estate agencies. The latter play an important role with regard to the acquisition and management of real estate on behalf of the city administration.

The success of the cooperation depends on the support of the different partners for the administrative approach, as well as the capacity they reserve for it. With regard to both these aspects, serious deficits were revealed in the course of the evaluation study (Huisman et al., 2005). The collation of relevant police information and fiscal information has turned out to be an important bottleneck for many of the subprojects. The authorization to collect this information apparently does not guarantee that the city authorities actually receive the requested data. Up until now, the various partners have predominantly paid lip service to the administrative approach, but have not integrated this strategy in their own working processes. The above-mentioned ambition to develop an integrated approach to organized crime has thus not yet been realised. In the successful handling of the Cosa Nostra in New York, the interaction between criminal investigations on the one hand and administrative measures on the other, were much better geared to each other (Fijnaut, 2002; Jacobs et al., 1999). In Amsterdam, instead of a twin-track approach, the criminal procedure and administrative prevention of organized crime are, with the exception of some successful operations, still separate and distinct programmes (Huisman et al., 2005).

Output and Outcome

Naturally, the implementation of an innovative project does take time. It was to be expected that the Van Traa project would encounter teething problems. Despite the conclusion that the administrative approach cannot yet be regarded as a form of integrated law enforcement, several results have been achieved. But to what extent do these results have a serious impact on organized crime in Amsterdam?

Literature on evaluation studies indicates the difficulty of determining the effects of policy measures, since it is often problematic to establish causality between measure and effect (Rossi, Freeman, & Lipsey, 1999). Reliable conclusions on causality require experimental research designs or statistical analysis. As criminal entrepreneurs generally try to conceal the illegal nature of their business, these methodological requirements cannot be met when evaluating the effects of measures on organized crime. With regard to preventative measures, the situation is even more complex. After all, it is difficult, if not impossible, to determine that something did *not* happen, due to the measures taken by the authorities. At most, only the plausibility that certain effects have been achieved can be established.

Because of the difficulties of requiring data and establishing causality between measure and effect, Levi and Maguire (2004) argue that for organized crime it remains largely a matter of belief that there is some effect. In the process of assessing the outcome of organized crime prevention, a distinction is therefore made between the *observed output* – the results of the efforts of the Van Traa team and its partners – and the plausible effects – including the unexpected and counterproductive ones – of this output on organized crime. This plausibility is based on the assumptions of the policy theory (Schoot, 2006). The term policy theory refers to a system of social and behavioural assumptions that underlie a public policy which have been reformulated in the form of propositions. These propositions reflect the beliefs of policymakers about the cognitions, attitudes, and behaviour of the policy's target groups: the people whom the policy is bound to affect. But they also refer to more structural factors about which policymakers have been making assumptions (Leeuw, 1991).

Observed Output: Results

During the period 1997–May 2004, 56 buildings were acquired by the city authorities and were given a bona fide exploitation. Four illegal casinos[1] and several establishments in the hotel and catering industries were closed down. About 20 licences for

---

[1] Since 1975 in the Netherlands, *Holland Casino* has been the only legal casino operator. *Holland Casino* is a foundation under government control which operates as a regular commercial company. A report published in 2000 (New round, new chances) proposed a revision of the government's gaming policy. It recommended that the embargo on new casinos should be lifted, with a phased introduction of a totally free market which would be open to any commercial operator able to meet the licensing requirements. These recommendations were not taken up because the government felt that the consequences were too unpredictable. Instead, in 2004, the government reaffirmed its restrictive gaming policy, stating that *Holland Casino* would remain the only legal licensed casino operator in the Netherlands.

bars and restaurants were refused or withdrawn, and eight permits in the catering industry were withdrawn. The most spectacular and far-reaching strategy was launched quite recently. Based upon the new BIBOB Act, the local authorities announced the withdrawal of the licences of the key players in the sex industry in the red-light district, unless the "entrepreneurs" concerned were able to submit a transparent accounting system. As none of the entrepreneurs was able to meet the desired standards of transparency, the licences for about hundred "windows" in the red light district were withdrawn at the end of November 2006. Despite the fact that a number of entrepreneurs decided to appeal to this decision, they may be forced to close their establishments in the near future.[2]

Plausible Effects on Organized Crime

As mentioned above, it is very difficult to assess the effects of preventive measures on organized crime. For several reasons, this is even more the case in the Van Traa project. Firstly, up until now no clear definition of organized crime has been used. In fact, in all official documents the term "organized" is placed between brackets. Secondly, no reliable analysis of the organized crime problem in the city exists. In order to be able to ascribe certain changes and effects to the Van Traa project, an in-depth analysis of the contemporary state of affairs is a *conditio sine qua non*. Neither the initial report by Fijnaut and Bovenkerk (1996)[3], nor later studies that reconfirmed the importance of Amsterdam as an international meeting place, safe haven and financial centre (Huisman, Huikeshoven, & van de Bunt, 2003; Unger, 2006)[4] have proved sufficiently specific or reliable for this purpose. Thirdly, in the policy theory that lies at the root of the administrative approach, many questions remain unanswered. What actions in particular are expected to weaken the position and influence of organized crime? What is the strength of the project, the withdrawal of permits or the possibility that criminals start investing their illegally

---

[2] One of these key players was supposedly on the list of the 16 "criminal entrepreneurs" Fijnaut and Bovenkerk referred to 10 years ago in their analysis of the vulnerability of the red-light district (Fijnaut & Bovenkerk, 1996).

[3] Although this analysis was a major incentive for the Van Traa project, the Van Traa team never received the original criminal intelligence analysis on which the report by Fijnaut and Bovenkerk was based. A local crime reporter released the names for the 16 anonymous "criminal" individuals described by Fijnaut and Bovenkerk and this information was used without being confirmed. Furthermore, one of the authors and even the criminal intelligence officers who made the analysis later questioned the reliability of the analysis.

[4] In a recent study for the Dutch ministry of Finance, Unger (2006) tried to estimate the extent of money laundering activities in the Netherlands. According to her calculations, money laundering amounts to 3.8 billion euros of "Dutch criminal revenues" and an additional inflow of 21 billion euro from crime abroad. Unger's report was severely criticized by various criminologists, who claimed that most of the assumptions that lie at the root of the calculations are highly questionable.

obtained income elsewhere? What kind of counter measures may be expected from criminal networks to prevent their positions being affected?

An important assumption in the context of the administrative approach is that due to administrative backlashes, poor law enforcement and a lack of interest, public administration in Amsterdam has lost control of certain areas and economic branches, and has given ample opportunity to criminals to commit crimes and launder their money. The solutions to this problem seem to be rather straightforward: more accurate information, better cooperation between the various departments within the municipality, and better cooperation between the public administration, the police, and the tax authorities. These solutions are in line with the ideological assumption that public administration should not facilitate organized crime. But does this also mean that organized crime is effectively combated? The empirical basis for the suggestion in the policy plans that the improvement of administrative processes and the introduction of a multi-agency approach will actually decrease the power and impact of organized crime is rather small. According to the plans, many policymakers tend to subscribe to such a causal relationship, but no one is able or willing to specify its nature.

A second problem with the above-mentioned assumption is that it is based on foreign images of what organized crime actually is. Organized crime in the Dutch context does have a specific character. Unlike the situation in Italy and New York, racketeering activities are not a common feature of organized crime in the Netherlands. In New York City, the administrative approach was directed to the core business of the Cosa Nostra families in the construction industry, the unions, etc., but in the Netherlands most forms of organized crime boil down to international smuggling activities. The production and trafficking of drugs, human trafficking, and human smuggling are the most important manifestations of this. The nature of organized crime in the Netherlands can thus be described as merely "transit crime" (Bunt, 2004; Kleemans, Brienen, & van de Bunt, 2002).[5]

Due to the fact that the vast majority of criminal activities take place in illegal markets (drugs, vice) and that most criminals are not interested whatsoever in generating economic and political power in the Netherlands, the image of a public administration which is facilitating organized crime has to be reconsidered. Naturally, in an *illegal* market, public administration by definition has no instruments to regulate the market as such. It is only when criminals abuse public

---

[5] At the moment, a criminal investigation is taking place concerning money laundering, extortion and contract killings in the real estate sector in Amsterdam. If the suspicions turn out to be justified, they would amount to a new development. Up until now, unlike in the USA and Italy, extortion has been regarded as an unfamiliar manifestation of organized crime in the Netherlands (apart from extortion within closed ethnic enclaves, such as the Chinese community). But even if the primary suspects of this alleged extortion case will be convicted, it is far too early to conclude that organized crime in the Netherlands is taking on American or Italian "traits." The focus of criminal entrepreneurs in the Netherlands is still on illegal markets, rather than wanting to control specific economic sectors.

facilities in the course of their activities in an illegal market or a money laundering operation, that national and local governments may have some options for counter-acting these activities.

In spite of the above-mentioned limitations, there are some indications of a plausible impact of the measures taken in the Van Traa project on organized crime. The files of the Van Traa team show several examples of entrepreneurs or certain business initiatives which could be related to organized crime and which were successfully barred from operating. Another clear example is the closure of the four known illegal casinos in the city. After the announcement that the Van Traa project would focus on the bar and catering industry in a certain run-down neighbourhood in the city, 25 establishments immediately ceased their business activities. The first internal evaluation of the implementation of the BIBOB Act showed that the possibility of screening discouraged several applicants from continuing the application process for licences. The withdrawal of 33 licences in the sex industry late November 2006 – concerning about 100 "windows" – on the basis of the BIBOB Act shows that the local authorities have put even more energy into preventing criminals from laundering their money in this branch.

Side Effects

The files of the Van Traa team also show some possible effects of displacement. There seem to be some geographical displacements: several entrepreneurs who were targeted in the city centre tried to set up businesses in other city districts and other cities. Other entrepreneurs remained active in the same area, but shifted to other types of business, in branches that are not subject to regulation or administrative control.

It is plausible that the Van Traa project also has had some counterproductive effects. So far, the start of a new subproject in a certain neighbourhood or branch has attracted a lot of media attention which has portrayed these neighbourhoods and branches as crime-ridden. This has had a stigmatising effect that is hard to overcome by positive news about the results of the project and it might have dis-couraged legitimate entrepreneurs from investing in such areas which are already vulnerable.

Residents and entrepreneurs in the red-light district[6] have also mentioned another possible counterproductive effect: the deterrent effect of the administrative approach seems to be stronger for small, marginal businesses than for larger, more powerful businesses who are better at knowing how to counteract the administrative lines of defence. This may lead to a situation in which relatively small businesses are taken over by larger, but not always flawless businesses. An indication of such an effect is

---

[6] In the course of the evaluation study, 16 entrepreneurs in the red-light district – some of them also living in the locality – were interviewed on their perceptions of the contemporary state of affairs in the area.

the concentration in the prostitution industry in the red-light district that has occurred since the introduction of a licensing system, the execution of the BIBOB Act and the operations of the Van Traa team. Many small brothel owners have sold their "windows" to a few larger players, who now own almost all the brothels in the area. Despite the above-mentioned fact that the licences for 33 brothels – mainly owned by these larger players – have been withdrawn on the basis of alleged connections to organized crime, the sixteen "key players" (or what remains of them: due to violent confrontations within the criminal world, more than half of them were killed in the last couple of years) are still present in the inner city. Their supposedly economic power has been challenged, but not broken.

The contemporary situation in the sex industry in the red-light district reflects an interesting paradox. On the one hand, this industry has officially been legalized since the legal ban on the exploitation of prostitution in the Netherlands was lifted in 2000.[7] On the other hand, entrepreneurs in this sector still encounter serious difficulties in finding a legitimate bank which is willing to support them financially. Financial institutions like to keep their distance from the sex industry, as they want to avoid being associated with "vice." According to the law, the exploitation of prostitution is thus a "normal and regular" economic activity, but the sector still operates in a moral twilight zone. Due to this paradox, entrepreneurs in the sex industry are still highly dependent on unofficial financial institutions and arrangements. As a result, the vulnerability of the sector of becoming involved in money laundering operations has not diminished since the process of legalization.[8] Parts of the sex industry have already gone "underground" and the transparency of the sector has not improved. A counterproductive effect of the increasing external pressure by the local authorities may be that more entrepreneurs will start looking for unofficial solutions to their problems. They may, for example, sell their regular brothels, sex houses and sex shops and become active in the escort branch, parts of which are not regulated by law and are not properly monitored. Up until now, the public reaction to these manifestations of displacement has been rather predictable: the authorities spend much time and energy in creating new opportunities to regulate, monitor and control all forms of behaviour that are considered to be undesirable.[9]

---

[7] According to the law that came into force on 1 October 2000, brothels where adult prostitutes choose to work voluntarily are no longer prohibited. At the same time, legislation on unacceptable forms of prostitution – human trafficking, minors – has become more severe. Before 2000, the exploitation of prostitution was officially a criminal offence, but the police and prosecution department paid little attention to this form of "crime." By legalizing the employment of prostitutes, the government is thought to exercise more control over the sex industry and counter abuse.

[8] In addition to involvement in money laundering activities, a recent evaluation study reveals that symbiotic relationships between human traffickers and representatives of the sex industry still exist (Daalder, 2007).

[9] New legislation and monitoring systems are being considered in order to regulate parts of the sex industry which have not yet been monitored and supervised by the state. Similarly, after several reports on unofficial value transfer systems in which money transfers were involved, the government suggested the introduction of new legal instruments to regulate money transfer systems.

According to the action-reaction principle, both legitimate and illegitimate entrepreneurs who want to avoid intensified state control will inevitably start looking for new opportunities in the unofficial economy. Due to this on-going "rat race," the transparency of markets will probably diminish, rather than increase.

Another paradox can be found in the gambling industry. As mentioned above, due to the multi-agency approach, the authorities have succeeded in closing down several casinos in the Netherlands – and in Amsterdam in particular – which were operating illegally. At the same time, the number of internet casinos has increased significantly during the last couple of years. Similar to the developments in the sex industry, the efforts of the authorities to put up administrative barriers in order to prevent organized crime controlling the gambling industry may also have had some counterproductive displacement effects. A *waterbed scenario* seems to have become manifest in which the authorities push hard on one side and criminal entrepreneurs start activities on the other.

Concluding Remarks

During the last decade, the authorities in Amsterdam have developed interesting new administrative instruments to prevent organized crime and disrupt markets which may be vulnerable for the intertwining of legal and illegal activities. It is much too early to analyse and draw any conclusions about the effects of the administrative approach. A decade or more is needed to find out whether this kind of innovative strategy has had any extra value in terms of diminishing the alleged power of organized crime. Even then, the impact will be hard to assess, because no clear target of the administrative approach was defined in advance and a solid, reliable threat analysis of the problem of organized crime in Amsterdam has not been carried out. Furthermore, simultaneously to the implementation of the administrative approach, other developments have taken place that may have influenced the local setting of organized crime and its containment. It is difficult, if not impossible, to isolate one specific project and to find empirical evidence of a causal relationship between a specific set of instruments used by the local authorities and the effects of these instruments on organized crime.

What we can conclude at this early stage is that some assumptions that lie at the root of the project are debatable, or have proven hard to employ in practice. The premise of the Van Traa project as an integrated multi-agency approach to organized crime has not yet been realised. The criminal, fiscal and administrative prevention of organized crime are, with the exception of some successful operations, still separate and distinct programmes. In terms of output and outcome, some indications of a plausible impact of the measures taken in the Van Traa project on serious forms of crime have been found. Some "dubious entrepreneurs" have successfully been barred from operating. The administrative approach – especially since the BIBOB Act was introduced – has seemed to offer some additional opportunities of putting up barriers against specific illegitimate business activities and entrepreneurs. However, we should keep in mind that administrative measures cannot bring about fundamental changes in the business climate in the branches concerned. In the sex and gambling industry some "rotten apples" may have been removed, but the basket is still contagious: the assumption that mala fide entrepreneurs can easily be replaced by bona fide businessmen demonstrates a rather naïve perception of the dynamics and modus operandi of these markets. By definition, in the "vice" industry a symbiotic relationship exists between legitimate and illegitimate activities.

Despite the above-mentioned deficits in the policy theory, many stakeholders are still convinced that the administrative approach is an appropriate remedy. As there is far less consensus on the evils at which the remedy is directed, there is a serious risk of net widening. The same development took place when the follow-the-money strategy in crime control was launched 15 years ago: the legal instruments that were developed in the Netherlands to deprive criminals of their illicit earnings were used in relatively minor cases, rather than in organized crime cases. Furthermore, the amount of money that criminals were actually deprived of during the first decade did not come close to the political expectations during the period that the legislation was drafted (Nelen, 2004). It may well be that in 10 years time the contemporary enthusiasm regarding the administrative approach will have evaporated and become transformed into sceptical conclusions, similar to the ones that Naylor formulated in relation to the follow-the-money strategies: "no one has been able to determine with any remote degree of confidence whether the proceeds-of-crime approach to crime control has had any discernible impact on the operation of illegal markets or on the amount, distribution, and behaviour of illegal income and wealth. The entire exercise rests on a series of inaccurate, or at least unprovable, assumptions..." (Naylor, 2004).

A more optimistic scenario is that in due time the partners involved will have succeeded in establishing an integrated multi-agency approach and will have been able to create more transparency in a number of economic sectors. The social sciences and criminology in particular, may also play a pivotal role in this respect. After all, we have to admit that our knowledge of the nature and extent of money laundering and the intertwinement of legitimate and illegitimate activities is still rather limited. Although scientific research cannot be expected to come up with ready-made solutions, criminological studies may contribute to the integral strategy for containing and preventing organized crime in various ways. First, in terms of the clarification and demystification of specific aspects of money laundering. Second, in terms of a critical analysis of the way in which the dual strategy is being conducted in daily practice. And, last but not least, in trying to provide greater insight into the effects – and side effects – of this strategy in the long run.

References

Bunt, H. G. van de. (2004). Organized crime policies in the Netherlands. In: C. Fijnaut & L. Paoli (Eds.), *Organized crime in Europe. Concepts, patterns and control policies in the European Union and beyond* (pp. 677–716). Dordrecht: Springer.

Daalder, A. L. (2007). *Prostitutie in Nederland na opheffing van het bordeelverbod*. Den Haag: Boom Juridische uitgevers, WODC, Onderzoek en Beleid, nr. 249.

Fijnaut, C. (Ed.) (2002). *The administrative approach to (organized) crime in Amsterdam*. Amsterdam: Public Order and Safety Department, City of Amsterdam.

Fijnaut, C., & Bovenkerk, F. (1996). *Inzake opsporing; enquête opsporingsmethoden, Bijlage XI: deelonderzoek IV onderzoeksgroep Fijnaut: De georganiseerde criminaliteit in Nederland: Een analyse van de situatie in Amsterdam – Een analyse van de situatie in Enschede, Nijmegen en Arnhem*. Den Haag: SDU, Tweede Kamer 1995–1996, 24072, nr. 20.

Huisman, W., & Nelen, H. (2007). Gotham unbound Dutch style, the administrative approach to organized crime in Amsterdam, *Crime, Law and Social Change*. (forthcoming)

Huisman, W., Huikeshoven, M., & Bunt, H. G. van de. (2003). *Marktplaats Amsterdam, Op zoek naar de zwakste schakel in de logistiek van criminele processen aan de hand van Amsterdamse rechercheonderzoeken*. Den Haag: Boom Juridische uitgevers.

Huisman, W., Huikeshoven, M., Nelen, H., Bunt, H. G. van de, & Struiksma, J. (2005). *Het Van Traa project. Evaluatie van de bestuurlijke aanpak van georganiseerde criminaliteit in Amsterdam*. Den Haag: Boom Juridische uitgevers.

Jacobs, J. B., Friel, C., & Radick, R. (1999). *Gotham unbound. How New York city was liberated from the grip of organized crime*. New York en London: New York University Press.

Kleemans, E. R., Brienen, M. E. I., & Bunt, H. G. van de. (2002). *Georganiseerde criminaliteit in Nederland, Tweede rapportage op basis van de WODC-monitor*. Den Haag: ministerie van Justitie, WODC, Onderzoek en beleid, nr. 198.

Leeuw, F. L. (1991). Policy theories, knowledge utilization and evaluation, knowledge and policy. *The International Journal of Knowledge Transfer, 4*(3), 73–91.

Levi, M., & Maguire, M. (2004). Reducing and preventing organized crime: An evidence-based critique. *Crime, Law and Social Change, 41*(5), 397–469.

Naylor, R. T. (2004). *Wages of crime; black markets, illegal finance, and the underworld economy* (revised ed.). Ithaca: Cornell University Press.

Nelen, H. (2004). Hit them where it hurts most; the proceeds-of-crime approach in the Netherlands. *Crime, Law and Social Change, 41*(5), 517–534.

Rossi, P. H., Freeman, H. E., & Lipsey, M. W. (1999). *Evaluation: A systematic approach*. Thousand Oaks: Sage Publications.

Schoot, C. R. A. van der. (2006). *Organized crime prevention in the Netherlands. Exposing the effectiveness of preventive measures*. Den Haag: Boom Juridische uitgevers.

Unger, B. (2006). De omvang en het effect van witwassen. *Justitiële verkenningen, 32*(2), 21–33.

About the Authors

Anton Blok studied cultural anthropology at the University of Amsterdam. His publications include *The Mafia of a Sicilian Village* (1974), *De Bokkerijders* (1991), and *Honour and Violence* (2001). He has taught at the universities of Amsterdam, Nijmegen, and Utrecht, and held visiting positions at the universities of Michigan, California (Berkeley), Yale, and Aix-en-Provence. He is now emeritus professor of cultural anthropology at the University of Amsterdam.

Stefano Becucci is an assistant professor at the University of Florence (Italy), Depatment at the University of Social Studies, His research interests are the process of migrants integration into the host societies, criminal markets and organized crime, the phenomenon of political participation and social movements against globalization among his most recent publications: *Globalizzasion e criminalità, Laterza, Roma-Bari, 2003 (co-author with M. Massari). Old and new actors in the Italian drug trade: ethnic reccession or functional specialization in European Journal on Crimianl policy and Research, 10, 2005.*

Tihomir Bezlov is Senior Analyst at the Center for the Study of Democracy (CSD) and Vitosha Research, where he has been working for the past 17 years. Mr. Bezlov has a wide-ranging expertise and has conducted over one hundred marketing and sociological studies. In the past 8 years he has held a leading role in CSD's research on corruption and crime, developing CSD's Corruption Assessment Index tool, National Crime Survey, and UNDP-Bulgaria's Early Warning Reports (1997–2002). He is the author of several studies on organized crime in Bulgaria, amongst which *The Drug Market in Bulgaria, Transportation, Smuggling and Organized Crime*, and *Corruption, Trafficking and Institutional Reform*.

Tim Boekhout van Solinge studied human geography at the University of Amsterdam and the University of Paris-Sorbonne, where he specialized in political geography and international relations. After finishing his masters, he studied drug use, drug trafficking and drug policy for over 10 years, which resulted in several books, articles and reports about drug control in France, the Netherlands, and Sweden. In 2004 he finished his Ph.D.: Dealing with Drugs in Europe (The Hague: Boom Legal Publishers) at Utrecht University. He now works as a lecturer in criminology at the Willem Pompe Institute for Criminal Law and Criminology, Utrecht University.

Since a few years his main research focus is shifting towards nature and environmental issues.

Henk van de Bunt graduated in law and sociology. He is professor of criminology at the Erasmus University Rotterdam. He was director of the Research and Documentation Center (WODC) of the Ministry of Justice (1994–2000) and professor of criminology at the Vrije Universiteit Amsterdam (1992–2005). His fields of interest are organized crime, corporate crime and the administration of criminal justice. He is one of the founders of the Centre for Information and Research on Organized Crime (CIROC) in The Netherlands. His recent publications include a.o. Van de Bunt and Kleemans (2007): Organised Crime in the Netherlands; The social organization of cannabis production (with Spapens & Rastovac, 2007).

Philip Gounev is a Research Fellow at the European Program of the Center for the Study of Democracy (CSD). Mr. Gounev is conducting research related to crime, arms-export controls, and discrimination issues. Prior to coming to CSD, Mr. Gounev completed a Master degree in International Security Policy at Columbia University's School of International and Public Affairs (New York, USA). Mr. Gounev also has extensive experience as finance professional with Putnam Investments and Beacon Health Strategies, both in Boston, MA. He has written over 20 publications. Recently he co-authored with Tihomir Bezlov *Crime Trends in Bulgaria 2000–2005 and Transportation, Smuggling and Organized Crime.*

James B. Jacobs is the Warren E. Burger Professor of Law at NYU School of Law, where he also serves as director of the law school's center for Research in Crime and Justice. A lawyer (J.D. University of Chicago, 1973) and a sociologist (Ph.D. University of Chicago, 1975), Professor Jacobs has been studying and writing about organized crime and organized crime control since the late 1980s. He has written four books on the general topic of the mafia's entrenchment in the U.S.'s political economy and on the government's quarter-century-long effort to eradicate this crime syndicate. The most recent book in this series is Mobsters, Unions and Feds: The Mafia and the American Labor Movement (NYU Press, 2006).

Wim Huisman is associate professor of criminology at the faculty of law of the VU University of Amsterdam. His main research interests are organizational crime and organised crime, and the regulatory and criminal justice policies aimed at these types of crime. Relevant publications include: Huisman, W. and P. Klerks, The role of public administration and the local businesses, in: Bunt, H.G. and C. van der Schoot, *Prevention of organised crime: A situational approach*, WODC, The Hague, 2003; Huisman, W., M. Huikeshoven, H. Nelen, H. van de Bunt en J. Struiksma, *Het Van Traa-project. Evaluatie van de bestuurlijke aanpak van georganiseerde criminaliteit in Amsterdam*, Boom Juridische uitgevers, Den Haag, 2005.

Anne-Marie Labalette received a master's degree from the School of Criminology, Université de Montréal. Her research deals with the impact of firearm legislation on violent crime trends. She has also worked as a parole officer with Correctional Services Canada and a research coordinator with juvenile correctional facilities in Quebec.

Francien Lankhorst graduated in law at the Vrije Universiteit Amsterdam, where she works as a lecturer in criminology. She is also working on her Ph.D. thesis on events that influence the opportunities for lawyers and notaries to help clients to commit crimes. She wrote in 2004 together with Nelen, "Professional Services and Organized Crime in the Netherlands," in: *Crime Law and Social Change*, and *Professionele dienstverlening en georganiseerde criminaliteit. Hedendaagse integriteitdilemma's van advocaten en notarissen*, Zeist: Kerckebosch bv.

Emanuel Marx is Professor Emeritus at Tel Aviv University. He is a social anthropologist who has done fieldwork among Bedouin of the Negev (Israel) and Bedouin of Sinai (Egypt), among North African immigrants in a new town in Israel and in Palestine refugee camps in territories occupied by Israel. His research interests are pastoral nomadism, violent behaviour, bureaucratic organizations and prehistoric man in the Levant. Recent writings include Nomads and Cities: The Development of a Conception. *In Shifts and Drifts in Nomad-Sedentary Relations*, Eds. Stefan Leder and Bernhard Streck (2005); The Political Economy of Middle Eastern and North African Pastoral Nomads. In *Nomadic Societies in the Middle East and North Africa: Entering the twenty-first Century*, Ed. Dawn Chatty (2006), Leiden: Brill.

Carlo Morselli is an associate professor at the School of Criminology, Université de Montréal. He is also a member of the International Centre for Comparative Criminology. He is the author of *Contacts, Opportunities, and Criminal Enterprise* (University of Toronto Press, 2005). His recent work has been in the fields of criminal networks, organized crime, and criminal achievement. Published work may be found in various journals, such as *Social Networks* (2007), *Global Crime* (2007), *Crime, Law, and Social Change* (2006), *Criminology* (2004, 2006), and *Critical Criminology* (2004).

Hans Nelen is a criminologist and has a law degree. Between 1986 and the beginning of 2001 he was employed as a senior researcher and research supervisor at the Research and Documentation Centre of the Ministry of Justice in the Netherlands (WODC), mainly involved in drug, fraud, organized crime, corporate crime and police research. Between 2001 and 2006 he was a senior lecturer and senior researcher at the Institute of Criminology of the Vrije Universiteit Amsterdam. Since January 1 2007 Nelen has been working as a professor of criminology at Maastricht University (UM). During the last decade Nelen published several books and articles on a variety of criminological subjects, i.e. corruption and fraud, dilemmas facing lawyers and notaries, the administrative approach to organized crime, the proceeds-of-crime approach, evaluation of legislation, and evaluation of law enforcement activities.

Antonio La Spina is full professor of sociology at the University of Palermo, department of Social sciences. He is chairman of the degree course of Communication Sciences and director of the School of Journalism "Mario Francese." He is scientific coordinator of the project "The costs of illegality" (Chinnici Foundation, Palermo), and former director of the targeted project on

Regulatory Impact Assessment (Ria) of the Department of Public Administration, (Presidency of the Council, Rome). He also worked on Ria for the Oecd, the Formez, the UE Presidency. Among his publications are *Mafia, legalità debole e sviluppo del Mezzogiorno*, Mulino, 2005; *La politica per il Mezzogiorno*, Mulino, 2003; The paradox of effectiveness: Growth, institutionalization and evaluation of anti-mafia policies in Italy. In C. Fijnaut & L. Paoli (Eds.) *Organised crime in Europe, Springer, 2004.*

Letizia Paoli is full professor at the Leuven Institute of Criminology of the Katholichs Universiteit Leuven Faculty of Law, Belgium. She received her Ph.D. in social and political sciences from the European University Institute in Florence in 1997. Since the 1990s she has published extensively on the Italian mafia, organised crime, drugs and illegal markets. Her most recent publications include *Mafia Brotherhoods: Organized Crime, Italian Style* (2003, Oxford University Press), *Organised Crime in Europe: Concepts, Patterns and Policies in the European Union and Beyond* (co-edited with Cyrille Fijnaut, 2004, Springer), and *Can the World Heroin Supply Be Reduced? What States Can and Cannot Accomplish in Drug Control* (with Peter Reuter and Victoria Greenfield, 2008, Oxford University Press).

Samuel L. Pineda is a program analyst in the Assistant Secretary's Office in the U.S. Department of State. He also served the Department of Defense as an Airborne Infantryman and a Military Intelligence contractor. Samuel received a Bachelor of Science degree in Criminal Justice and has been awarded the degree Master of Public Administration from San Diego State University where he conducted primary research on human trafficking activities in Tijuana, Mexico.

Dina Siegel is an assistant professor at the Department of Criminology and Criminal Law, Vrije Universiteit Amsterdam. She studied sociology and social anthropology at the Tel-Aviv University, Israel and obtained her Ph.D. in cultural anthropology at the VU University of Amsterdam, the Netherlands. She has studied and published several articles on post-Soviet organized crime, terrorism, women trafficking, criminality in diamond sector, and on XTC distribution and XTC policy in the Netherlands. She is one of the founders of the Centre for Information and Research on Organized Crime (CIROC). In 2003 she edited (together with Van de Bunt and Zaitch) *Global Organized Crime. Trends and Developments* (Kluwer). She conducted ethnographic research on Russian-speaking criminals in the Netherlands, *Russian biznes in the Netherlands* (2005, Meulenhoff).

Richard Staring studied Cultural Anthropology and is currently appointed at the Department of Criminology of Erasmus University Rotterdam, the Netherlands. He has published on poverty, crime, international and irregular migration flows, as well as on processes of incorporation of illegal immigrants. He recently published on the subject of human smuggling (Kerckebosch, 2005) and on criminal structures around the real estate sector (SWP, 2007). Staring is preparing a publication on the irregular migration processes of undocumented Turkish immigrants in the Netherlands (Berghahn, 2007) based on ethnographic fieldwork in the Netherlands and Turkey.

Dave Tanguay received a master's degree from the School of Criminology, Université de Montréal. His research deals with criminal homicide and account settlement trends. He is currently a crime analyst with the Royal Canadian Mounted Police.

Sheldon X. Zhang is Professor of Sociology at San Diego State University in California. He received his Ph.D. from the University of Southern California in 1993. His main research interests revolve around transnational organized crime and community corrections. His publications have appeared in such journals as *Criminology, British Journal of Criminology, Research in Crime and Delinquency*, and *Crime & Delinquency*. He is also the co-author (with Robert Winslow) of a new text: Criminology – A Global Perspective (Prentice Hall).

Appendix: Scientific Board

The Scientific Board:
Prof. Dr. Jay Albanese (Virginia Commonwealth University, Richmond, Virginia, USA)
Prof. Dr. Mike Levi (Cardiff University, Wales, UK)
Prof. Dr. Nikos Passas (Northeastern University, Boston, USA)
Prof. Dr. Ernesto Savona (Università Cattolica di Milano, Italy)
Prof. Dr. Menachem Amir (Hebrew University, Jerusalem, Israel)
Prof. Dr. Henk van de Bunt (Erasmus University Rotterdam, The Netherlands)
Dr. Andrea diNicola (Università di Trento, Italy)
Dr. Edward Kleemans (WODC, Ministry of Justice, The Hague, The Netherlands)
Dr. Klaus von Lampe (Freie Universität Berlin, Germany)

Index